T0140810

Martin Löhning

From Model Reduction to Efficient Predictive Control with Guarantees

Logos Verlag Berlin

λογος

Bibliografische Information der Deutschen Nationalbibliothek

Die Deutsche Nationalbibliothek verzeichnet diese Publikation in der Deutschen Nationalbibliografie; detaillierte bibliografische Daten sind im Internet über http://dnb.d-nb.de abrufbar.

D 93

ISBN 978-3-8325-5485-9

Logos Verlag Berlin GmbH
Georg-Knorr-Str. 4, Geb. 10,
D-12681 Berlin
Germany

Tel.: +49 (0)30 / 42 85 10 90
Fax: +49 (0)30 / 42 85 10 92
http://www.logos-verlag.de

Acknowledgements

This thesis presents the results of my research at the Institute for Systems Theory and Automatic Control at the University of Stuttgart. Many people supported me during this time as a research and teaching assistant and afterwards, while writing this thesis.

First of all, I want to express my greatest appreciation to my adviser Prof. Dr.-Ing. Frank Allgöwer. Most importantly, he provided an inspiring and international research environment with numerous opportunities and impressive young researchers. For example, he gave me the opportunity to learn considerable more on control theory by visiting several lectures and a summer school. Moreover, I profoundly thank him for his invaluable advice, continuing encouragement, and patience.

I also express my gratitude to Prof. Dr. Bernard Haasdonk. His lectures and research enriched my understanding of model reduction. Notably, he acquainted me with the a-posteriori error bounds which are an important basis for the key results in this thesis. I am sincerely grateful for his supportive statements which helped me finish this thesis. Furthermore, I would like to thank Prof. Dr.-Ing. habil. Boris Lohmann and Prof. Dr.-Ing. Oliver Riedel for their interest in my work and being members of my doctoral examination committee.

The time at the institute would not have been so worthwhile and enjoyable without all the colleagues. I thank all members of the institute for the open atmosphere and the interesting discussions. Particularly, I would like to acknowledge my appreciation to my office mates and coauthors Ulrich Münz, Marcus Reble, Matthias Lorenzen, Jan Hasenauer, and Shuyou Yu. I am particularly grateful to Jan Hasenauer for his crucial stimulations and Marcus Reble for the all-embracing discussions, his critical questions and constructive comments. Both have been great instructors and true friends.

Furthermore, Matthias Löhning, Marcus Reble, Matthias Hirche, Matthias Lorenzen, and Matthias Müller improved the thesis by proofreading. I would like to thank all of them for their valuable comments.

Finally, I am deeply grateful to my wife Angelika and my children Emilia and Ferdinand for their persistent patience. Furthermore, I would like to express my deep gratitude to my parents Gerlinde and Wilfried for their admirable support and love during my entire life.

Leonberg, July 2021
Martin Löhning

Table of Contents

List of Figures

List of Tables

List of Abbreviations

Abbreviation	Description	Page
ARE	algebraic Riccati equation	77
DAE	differential algebraic equation	5
FPU	Fermi-Pasta-Ulam	54
I/O	input/output	5
LQR	linear-quadratic regulator	76
LTI	linear time invariant	2
MAPK	mitogen-activated protein kinase	14
MPC	model predictive control	1
ODE	ordinary differential equation	2
PDE	partial differential equation	5
$\mathcal{P}$-MPC	model predictive control using the plant model given by Algorithm 2.8	26
POD	proper orthogonal decomposition	20
QCQP	quadratically constrained quadratic program .	118
$\mathcal{R}$-MPC	model predictive control using the reduced model given by Algorithm 2.14	29
SOCP	second-order cone program	16
SVD	singular value decomposition	21
Δ-MPC	model predictive control using the reduced model and error bounding system given by Algorithm 5.9	85

List of Symbols

The list of symbols is divided into, first, mathematical symbols, second, system variables and control parameters, and, third, tubular reactor variables and parameters.

The symbols are sorted by numbers, the Roman alphabet, the Greek alphabet, and further symbols.

Symbols appearing only in one section are defined locally. If confusion with the notation is improbable, then symbols can have a meaning depending on the context. For example, the symbol C is used, first, to describe the polytopic constraints within the MPC sections, second, to denote the reactant concentration of the tubular reactor, and, third, in the state space notation in Section 2.1.1.

In general, uppercase letters denote matrices and lowercase letters refer to scalars and vectors. In order to follow the common notation, exceptions are made within the system variables and control parameters as well as the tubular reactor variables and parameters.

Mathematical Symbols

Symbol	Explanation	Page
Sets		
$\mathcal{L}_2^n$	Lebesgue space of square integrable functions $f : \mathbb{R}_{0+} \to \mathbb{R}^n$	17
$\mathcal{L}_\infty^n[a, b]$	Lebesgue space of functions $f : [a, b] \subset \mathbb{R}_{0+} \to \mathbb{R}^n$ that are bounded on $[a, b]$ with $a < b$	37
$\mathcal{PC}_{[a,\,b]}^n$	set of all piecewise continuous functions $f : [a, b] \subset \mathbb{R} \to \mathbb{R}^n$ with $a < b$	25
$\mathbb{N}_0$	set of nonnegative integers	25
$\mathbb{R}$	set of real numbers	17
$\mathbb{R}_{0+}$	set of nonnegative real numbers	17
$\mathbb{R}_{++}$	set of positive real numbers	38
$\mathbb{S}_{0+}^n$	set of positive semidefinite and symmetric matrices in $\mathbb{R}^{n\times n}$	63
$\mathbb{S}_{++}^n$	set of positive definite and symmetric matrices in $\mathbb{R}^{n\times n}$	50
$\left\{x^{(i)}\right\}_{i=1}^n$	set with n possibly repeated elements $x^{(i)}$. The order in which the elements are listed is of importance.	20
$[a, b]$	closed interval $\{t \in \mathbb{R} \vert a \leq t \leq b\}$ with $a, b \in \mathbb{R}$	18

Functions

$\mathrm{diag}(A_1, \ldots, A_n)$	block-diagonal matrix with the matrices $A_1, \ldots, A_n$ on the diagonal	36
$\mathrm{diag}(x_1, \ldots, x_n)$	matrix with the scalars $x_1, \ldots, x_n$ on the diagonal	85
e^A	matrix exponential for $A \in \mathbb{R}^{n \times n}$, i.e., $\mathrm{e}^A = \sum_{i=0}^{\infty} \frac{1}{i!} A^i$	21
e^x	exponential function for $x \in \mathbb{R}$ in which e is Euler's number	63
$\exp(x)$	exponential function e^x of $x \in \mathbb{R}$ in which e is Euler's number	32
$\ln(x)$	logarithm with base e for $x \in \mathbb{R}_{++}$	100
$\sup(x)$	supremum of $x \in \mathbb{R}$	67
$\mathrm{trace}(A)$	trace of the matrix $A \in \mathbb{R}^{n \times n}$	52
$\lambda_{\max}(A)$	largest real part of the eigenvalues of the matrix $A \in \mathbb{R}^{n \times n}$	61
$\lambda_{\min}(A)$	smallest real part of the eigenvalues of the matrix $A \in \mathbb{R}^{n \times n}$	63
$\sigma_{\max}(A)$	largest singular value of the matrix $A \in \mathbb{R}^{n \times m}$	63
$\sigma_{\min}(A)$	smallest singular value of the matrix $A \in \mathbb{R}^{n \times n}$	63
$\lvert x \rvert$	absolute value of $x \in \mathbb{R}$	46
$\lVert x \rVert_1$	absolute norm of $x \in \mathbb{R}^n$, i.e., $\lVert x \rVert_1 = \sum_{i=1}^{n} \lvert x_i \rvert$	43
$\lVert x \rVert$	Euclidean norm of $x \in \mathbb{R}^n$, i.e., $\lVert x \rVert = \sqrt{x^\mathsf{T} x}$	21
$\lVert x \rVert_G$	weighted euclidean norm of $x \in \mathbb{R}^n$, i.e., $\lVert x \rVert_G = \sqrt{x^\mathsf{T} G x}$ for $G \succ 0$	61
$\lVert A \rVert$	spectral norm of $A \in \mathbb{R}^{n \times m}$, i.e., $\lVert A \rVert = \sigma_{\max}(A)$	21
$\lVert x(\cdot) \rVert_{\mathcal{L}_2}$	$\mathcal{L}_2$-norm of $x(\cdot) \in \mathcal{L}_2^n$, i.e., $\lVert x(\cdot) \rVert_{\mathcal{L}_2} = \left(\int_0^\infty x^\mathsf{T}(t) x(t)\, \mathrm{d}t\right)^{1/2}$	96
$\kappa(A)$	condition number of the matrix A, i.e., $\kappa(A) = \lVert A \rVert \lVert A^{-1} \rVert$	63
$\mathrm{d}f(x_0)/\,\mathrm{d}x$	derivative of f with respect to x at x_0	41
$\partial f(x_0)/\partial x_i$	partial derivative of f with respect to x_i at x_0	32
$x > y$	$x - y \in \mathbb{R}^n$ positive, i.e., $x_i > y_i$ for all $i = 1, \ldots, n$	18
$x \geq y$	$x - y \in \mathbb{R}^n$ nonnegative, i.e., $x_i \geq y_i$ for all $i = 1, \ldots, n$	21
$x < y$	$x - y \in \mathbb{R}^n$ negative, i.e., $x_i < y_i$ for all $i = 1, \ldots, n$	20
$x \leq y$	$x - y \in \mathbb{R}^n$ nonpositive, i.e., $x_i \leq y_i$ for all $i = 1, \ldots, n$	21
$A \prec 0$	matrix A is negative definite	50
$A \preceq 0$	matrix A is negative semidefinite	63
$A \succ 0$	matrix A is positive definite	50
$A \succeq 0$	matrix A is positive semidefinite	52

Accents

$\dot{f}(t, x)$	derivative of f with respect to the time t	17

$\ddot{f}(t,x)$	second derivative of f with respect to the time t . .	46
$\dddot{f}(t,x)$	third derivative of f with respect to the time t . . .	46

Superscripts

$A^{1/2}$	unique square root of the matrix $A \in \mathbb{S}^n_{0+}$	51
A^{-1}	inverse of the matrix $A \in \mathbb{R}^{n\times n}$	18
A^{T}	transpose of the matrix $A \in \mathbb{R}^{n\times m}$	18
$x^{(i)}$	i-th element of a set $\{x^{(i)}\}_{i=1}^{N}$ with possibly repeated elements	20

Miscellaneous

0_n	$n \times 1$ vector of zeros	50
$0_{n\times m}$	$n \times m$ matrix whose entries are all 0	50
$1_{n\times m}$	$n \times m$ matrix whose entries are all 1	131
I_n	$n \times n$ identity matrix	18
s	Laplace variable	45
$\Phi(t \mid a, b)$ . . .	probability density of the uniform distribution in the interval $t \in [a,b] \subset \mathbb{R}$	45
e	Euler's number	21
$\in$	is element of	17
$\gg$	much greater than	40
$\jmath$	imaginary number, i.e., $\jmath = \sqrt{-1}$	55
$(x_1, \ldots, x_n)$.	tuple with n ordered components	49
$\binom{n}{k}$	binomial coefficient, i.e., $\binom{n}{k} = (n!)/\big(k!(n-k)!\big)$ with $n, k \in \mathbb{N}_{0+}$	45
;	the semicolon within the argument of a function $f(t;p)$ separates the variable t from the parameters p . . .	17
$z^{(i)} \sim p_z(z)$ . .	the sample member $z^{(i)}$ is drawn according to $p_z(z)$	40

System Variables and Control Parameters

Symbol	Explanation	Page
Accents		
$\bar{x}(\cdot; t_i)$	signal predicted within an MPC scheme at sampling instant t_i	25
Superscripts		
$x^*(\cdot; t_i)$	signal which solves the optimization problem corresponding to the MPC scheme at the sampling instant t_i	25

Subscripts

x_{P}	variable corresponding to the plant	23
$x_{\mathcal{P}}$	variable corresponding to $\mathcal{P}$-MPC, i.e, the plant model is used for the prediction	25
x_{R}	variable corresponding to the reduced model	18
$x_{\mathcal{R}}$	variable corresponding to $\mathcal{R}$-MPC, i.e, the reduced model is used for the prediction	29
y_{SS}	steady state output of the detailed model for the constant input u_{SS}	39
x_{Δ}	variable corresponding to Δ-MPC, i.e, the reduced model and error bounding system is used for the prediction	84

Miscellaneous

A	state matrix of the preprocessed model	20
A_{P}	state matrix of the plant	60
A_{R}	state matrix of the reduced model	20
$\mathcal{A}^{(s)}$	state matrix of the Jacobi linearization of the reduced model at the steady state $\left(u_{\mathrm{SS}}^{(s)}, y_{\mathrm{SS}}^{(s)}\right)$	50
B	input matrix of the preprocessed model	20
B_{P}	input matrix of the plant	60
B_{R}	input matrix of the reduced model	20
C	matrix within the polytopic constraints for the state and input of the preprocessed model	79
C_{P}	matrix within the polytopic constraints for the state and input of the plant	23
d	vector within the polytopic constraints for the state and input of the preprocessed model	79
d_{P}	vector within the polytopic constraints for the state and input of the plant	23
E	integrated squared output error	39
$E_{\mathcal{P}}$	terminal cost within $\mathcal{P}$-MPC	27
E_{Δ}	terminal cost within Δ-MPC	93
F	quadratic stage cost of the preprocessed model	80
f	vector field of the state of the detailed model	17
$\check{F}$	equivalent stage cost for the preprocessed model when using the LQR for prestabilization	88
F_{P}	quadratic stage cost of the plant	24
f_{P}	vector field for the state of the plant	23
F_{R}	quadratic stage cost of the reduced model	81
f_{R}	vector field of the state of the reduced model	39
$\mathcal{F}_{\mathcal{P}}$	region of attraction of the plant in closed loop with $\mathcal{P}$-MPC	27

$\mathcal{F}_{\Delta}$	region of attraction of the plant in closed loop with Δ-MPC	99
g_{R}	vector field mapping the initial condition of the detailed model to the initial condition of the reduced model	39
h	vector field of the output of the detailed model	17
h_{R}	vector field of the output of the reduced model	39
$\mathcal{I}$	subspace of $\mathbb{R}^{n_{\mathrm{R}}}$ containing the origin, the projection of the state of the preprocessed model, and the predicted state of the reduced model	83
J	cost functional	39
$\hat{J}$	approximated cost functional	40
$\hat{J}_{\mathrm{m}}$	modified cost function based on the ansatz functions $m(\xi, \mu)$	42
$J_{\mathcal{P}}$	finite horizon cost functional of $\mathcal{P}$-MPC	25
$J_{\mathcal{R}}$	finite horizon cost functional of $\mathcal{R}$-MPC	29
J_{Δ}	finite horizon cost functional of Δ-MPC	83
J^{inf}	infinite horizon cost functional for the preprocessed model	80
$\check{J}^{\mathrm{inf}}$	equivalent infinite horizon cost functional for the preprocessed model when using the LQR for prestabilization	102
$J_{\mathrm{P}}^{\mathrm{inf}}$	infinite horizon cost functional for the plant	24
$J_{\mathcal{P}}^{\mathrm{inf}}$	infinite horizon cost functional for the plant in closed loop with $\mathcal{P}$-MPC	140
$J_{\mathcal{R}}^{\mathrm{inf}}$	infinite horizon cost functional for the plant in closed loop with $\mathcal{R}$-MPC	140
$J_{\Delta}^{\mathrm{inf}}$	infinite horizon cost functional for the plant in closed loop with Δ-MPC	140
K_{P}	feedback matrix for the prestabilization of the plant	61
K_{Ω}	feedback matrix for the terminal control law	93
m	vector of ansatz functions	41
n	order of the plant, preprocessed, and detailed model	17
n_C	number of constraints for the state and input	23
n_{R}	order of the reduced model	18
n_S	number of steady states	39
n_T	number of trajectories	40
n_{U}	dimension of the input	17
n_{Y}	dimension of the output	17
n_{θ}	dimension of the parameter vector θ which weights the ansatz functions m	41
n_{φ}	dimension of the parameter vector φ for the input	38
$P_{\mathrm{P}}^{\mathrm{ARE}}$	solution $P_{\mathrm{P}}^{\mathrm{ARE}} \in \mathbb{S}_{0+}^{n}$ of the algebraic Riccati equation for the plant	87

Symbol	Description	Page
$P_{\mathrm{R}}^{\mathrm{ARE}}$	solution $P_{\mathrm{R}}^{\mathrm{ARE}} \in \mathbb{S}_{0+}^{n_{\mathrm{R}}}$ of the algebraic Riccati equation for the reduced model	94
$p_{\mathrm{x},\varphi}$	weighting function for the initial condition and the input parameterized by φ	38
P_{Δ}^{Ω}	matrix $P_{\Delta}^{\Omega} \in \mathbb{S}_{++}^{n_{\mathrm{R}}}$ in the terminal set of Δ-MPC	93
Q	matrix $Q \in \mathbb{S}_{0+}^{n}$ penalizing the state of the preprocessed model in the quadratic stage cost	80
Q_{P}	matrix $Q_{\mathrm{P}} \in \mathbb{S}_{0+}^{n}$ penalizing the state of the plant in the quadratic stage cost	24
$Q_{\mathcal{P}}^{\Omega}$	matrix $Q_{\mathrm{P}}^{\Omega} \in \mathbb{S}_{0+}^{n}$ penalizing the terminal state in $\mathcal{P}$-MPC	27
Q_{R}	matrix $Q_{\mathrm{R}} \in \mathbb{S}_{0+}^{n_{\mathrm{R}}}$ penalizing the state of the reduced model in the quadratic stage cost	81
Q_{Δ}^{Ω}	matrix $Q_{\Delta}^{\Omega} \in \mathbb{S}_{0+}^{n_{\mathrm{R}}}$ penalizing the terminal state in Δ-MPC	93
R	matrix $R \in \mathbb{S}_{0+}^{n_{\mathrm{U}}}$ penalizing the input of the preprocessed model in the quadratic stage cost	80
r	residual of model reduction by projection	21
R_{P}	matrix $R_{\mathrm{P}} \in \mathbb{S}_{0+}^{n_{\mathrm{U}}}$ penalizing the input of the plant in the quadratic stage cost	24
S	matrix $S \in \mathbb{R}^{n \times n_{\mathrm{U}}}$ penalizing the cross term between the state and input of the preprocessed model in the quadratic stage cost	80
S_{R}	matrix $S_{\mathrm{R}} \in \mathbb{R}^{n_{\mathrm{R}} \times n_{\mathrm{U}}}$ penalizing the cross term between the state and input of the reduced model in the quadratic stage cost	81
$\mathcal{S}$	set of locally exponentially stable steady states	39
T	prediction horizon	25
t	time	17
T_{end}	time until a good approximation of the I/O behavior is desired	38
t_i	sampling instant $t_i = i\delta$, $i \in \mathbb{N}_0$	25
T_{P}	state transformation matrix for the preprocessing of the plant with $P \succ 0$	62
$\mathcal{T}$	sample of points from I/O trajectories	40
u	input of the preprocessed or detailed model	17
u_{P}	input of the plant	23
$u_{\mathcal{P}}$	input of the plant resulting from $\mathcal{P}$-MPC	26
$u_{\mathcal{R}}$	input of the plant resulting from $\mathcal{R}$-MPC	29
u_{Δ}	input of the plant resulting from Δ-MPC	85
u_{φ}	input of the detailed model parameterized by $\varphi \in \mathbb{R}^{n_{\varphi}}$	38
V	full-column rank matrix used for model reduction by projection	18

Symbol	Description	Page
W	full-column rank matrix used for model reduction by projection	18
X	symmetric matrix of dimension n_{R}	50
x	state vector of the preprocessed or detailed model	17
x_0	initial condition of the preprocessed or detailed model	17
x_{P}	state of the plant	23
$x_{\mathrm{P},0}$	initial condition of the plant	23
x_{R}	state of the reduced model	18
$\mathcal{X}$	set of possible states of the preprocessed model given the state of the reduced model and the error bound	65
Y	symmetric matrix of dimension n_{R}	52
y_{D}	output of the detailed model	17
y_{R}	output of the reduced model	18
α	scaling factor within the bound of the matrix exponential	63
β	decay rate within the bound of the matrix exponential	63
γ_{x}	parameter in the terminal set of Δ-MPC	93
γ_{Δ}	parameter in the terminal set of Δ-MPC	93
Δ	error bound	64
δ	sampling time $\delta < 0$	25
θ	parameter vector θ which weights the ansatz functions m	41
μ	extended input $\mu = \left[u_\varphi \ \dot{u}_\varphi \ \ldots \ \frac{\mathrm{d}^{n_{\mathrm{R}}}}{\mathrm{d}t^{n_{\mathrm{R}}}} u_\varphi\right]^{\mathsf{T}}$	40
$\mu^{(i)}$	extended input μ evaluated at $t^{(i)}$, $x_0^{(i)}$, $u_\varphi^{(i)}$	42
ν	function of the n_{R}-th state of the reduced model in the observability normal form	40
ξ	state of the reduced model in observability normal form, i.e., $\xi = \left[y_{\mathrm{R}}, \dot{y}_{\mathrm{R}}, \ldots, \frac{\mathrm{d}^{n_{\mathrm{R}}-1}}{\mathrm{d}t^{n_{\mathrm{R}}-1}} y_{\mathrm{R}}\right]^{\mathsf{T}}$	40
$\xi_{\mathrm{D}}^{(i)}$	vector of the outputs of the detailed model up to the $(n_{\mathrm{R}}-1)$-th derivative evaluated at $t^{(i)}$, $x_0^{(i)}$, $u_\varphi^{(i)}$, i.e., $\xi_{\mathrm{D}}^{(i)} = \left[y_{\mathrm{D}}^{(i)}\left(t^{(i)}, x_0^{(i)}, u_\varphi^{(i)}\right), \ldots, \frac{\mathrm{d}^{n_{\mathrm{R}}-1}}{\mathrm{d}t^{n_{\mathrm{R}}-1}} y_{\mathrm{D}}^{(i)}\left(t^{(i)}, x_0^{(i)}, u_\varphi^{(i)}\right)\right]^{\mathsf{T}}$	42
ρ	scalar and real parameter	51
ϱ	upper bound for the norm of the residual r scaled by α	115
Σ_{D}	detailed model	17
Σ_{P}	plant model	23
Σ_{PP}	preprocessed model	62
Σ_{R}	reduced model	18
Σ_{Δ}	error bounding system	64
φ	parameter vector of the input	38
$\Omega_{\mathcal{P}}$	terminal set within $\mathcal{P}$-MPC	25
Ω_{Δ}	terminal set within Δ-MPC	93

Tubular Reactor Variables and Parameters

Symbol	Explanation	Page
C [mol/l]	reactant concentration	31
C_0 [mol/l]	reactant concentration at time $t = 0$	32
C_{in} [mol/l]	inlet reactant concentration	31
C_{lin} [mol/l]	reactant concentration used for the linearization of the PDE	129
C_{nom} [mol/l]	nominal inlet reactant concentration	31
C_{out} [mol/l]	outlet reactant concentration	31
C_{P} [cal/kg·K]	specific heat at constant pressure	32
d [m]	reactor diameter	32
E [cal/mol]	activation energy	32
G_{r} [l·K/mol·s]	abbreviation for $\frac{-\Delta H\, k_0}{\rho\, C_{\text{P}}}$	32
h [W/m²·K]	heat transfer coefficient	32
H_{r} [1/s]	abbreviation for $\frac{4\, h}{d\, \rho\, C_{\text{P}}}$	32
k_0 [1/s]	kinetic constant	32
L [m]	reactor length	32
P	product	30
R	reactant	30
R_{gas} [cal/mol·K]	ideal gas constant	32
T [K]	fluid temperature	31
T_0 [K]	fluid temperature at time $t = 0$	32
T_{in} [K]	inlet fluid temperature	31
T_{J} [K]	temperatures of the three jackets	31
$T_{\text{J,lin}}$ [K]	temperatures of the three jackets used for the linearization of the PDE	129
T_{lin} [K]	fluid temperature used for the linearization of the PDE	129
T_{nom} [K]	nominal inlet fluid temperature	30
T_{out} [K]	outlet fluid temperature	31
T_{w} [K]	reactor wall temperature	31
v [m/s]	fluid superficial velocity	32
z [m]	axial coordinate	31
ΔH [cal/mol]	heat of reaction	32
Δz [m]	step size of spatial discretization	32
ρ [kg/m³]	fluid density	32

Abstract

This thesis presents methods that use model reduction to reduce the complexity of a dynamical model to a practicable level. In addition to procedures for model reduction, we introduce algorithms for bounding the error between the detailed and reduced model as well as predictive control using reduced models for prediction. All approaches provide guarantees either for the reduced model or while using the reduced model. At the same time, we mind the computational tractability of the procedures. The applicability of the proposed methods is demonstrated by means of a nonisothermal tubular chemical reactor.

The first contribution is a model reduction procedure to approximate the input-output map of continuous-time nonlinear ordinary differential equations. The reduced model is parameterized with the observability normal form. Using a sample of simulated input-output trajectories, the parameters are computed by convex optimization. A low complexity of the functional expression is promoted by sparsity enhancing ℓ_1-minimization. In addition, we extend the method to preserve the location and local exponential stability of multiple steady states.

Furthermore, we improve an existing a-posteriori bound of the model reduction error for linear models. The generalized error bound is given by an asymptotically stable scalar ordinary differential equation, which results, in general, in a considerably tighter bound with a comparable computational demand.

Finally, we propose a novel model predictive control scheme using reduced models for linear time-invariant systems. This model predictive control scheme uses the developed bound of the model reduction error to guarantee asymptotic stability as well as satisfaction of hard input and state constraints despite the error between the reduced model used for the prediction and the high-dimensional plant. Moreover, we show that the proposed model predictive control scheme minimizes the infinite horizon cost functional for the plant for a common choice of design parameters. In this case, the proposed scheme ensures an upper bound for the cost functional value of the closed loop with the detailed plant model despite using a reduced model for the prediction. For discrete-time plants we show that the optimization problem of the model predictive control scheme can be reformulated as a second-order cone program.

Deutsche Kurzfassung

In dieser Arbeit werden verschiedene Methoden entwickelt, die durch den Einsatz von Modellreduktion die Komplexität eines dynamischen Modells beherrschbar machen. Neben einem Verfahren für die Modellreduktion stellen wir sowohl einen Algorithmus für die Beschränkung des Fehlers zwischen dem detaillierten und dem reduzierten Modell als auch eine Methode für die modellprädiktive Regelung basierend auf einem reduzierten Modell vor. Für alle Ansätze leiten wir Garantien entweder für das reduzierte Modell oder für die Verwendung des reduzierten Modells her. Gleichzeitig achten wir darauf, dass die entwickelten Algorithmen mit derzeitigen Computern gelöst werden können. Die Anwendbarkeit der Methoden wird anhand eines nicht-isothermen chemischen Rohrreaktors demonstriert.

Als erste Methode schlagen wir ein Verfahren vor, das das Ein-/Ausgangsverhalten einer zeitkontinuierlichen nichtlinearen gewöhnlichen Differentialgleichung approximiert. Das reduzierte Modell wird mit der Beobachternormalform parametrisiert. Basierend auf einer Stichprobe von simulierten Ein-/Ausgangstrajektorien sind die Parameter durch die Lösung eines konvexen Optimierungsproblems bestimmt. Eine geringe Komplexität der Funktion des reduzierten Modells wird durch eine ℓ_1-Minimierung erreicht. Darüber hinaus erweitern wir die Methode, so dass die Position und lokale exponentielle Stabilität der stationären Zustände erhalten bleiben.

Außerdem verbessern wir eine bestehende a-posteriori Schranke für den Modellreduktionsfehler für lineare Modelle. Die verallgemeinerte Fehlerschranke wird durch eine asymptotisch stabile skalare gewöhnliche Differentialgleichung beschrieben und führt, im Allgemeinen, zu einer deutlich genaueren Schranke mit einem vergleichbaren Rechenaufwand.

Weiterhin stellen wir ein neuartiges modellprädiktives Regelungsverfahren basierend auf reduzierten Prädiktionsmodellen für lineare zeitinvariante Systeme vor. Das entwickelte Verfahren verwendet die verallgemeinerte Schranke für den Modellreduktionsfehler, um asymptotische Stabilität sowie die Einhaltung von Eingangs- und Zustandsbeschränkungen trotz der Abweichung zwischen dem reduzierten Prädiktionsmodell und dem detaillierten Streckenmodell zu garantieren. Zudem beweisen wir für eine übliche Wahl von Entwurfsparametern, dass das modellprädiktive Regelungsverfahren, trotz des reduzierten Prädiktionsmodells, das Kostenfunktional für die Strecke minimiert und eine obere Schranke für den Wert des Kostenfunktionals des geschlossenen Kreises mit der Strecke gewährleistet. Für zeitdiskrete Strecken zeigen wir, dass das Optimierungsproblem des modellprädiktiven Regelungsverfahrens in ein Second-Order Cone Programm umgeformt werden kann.

Chapter 1

Introduction

1.1 Motivation

In the industrial environment, model predictive control (MPC) is getting more and more popular. Two reasons are the increasing computational power and more demanding performance requirements for controlled systems. The latter results in more difficult control task, which render the well-known proportional-integral-derivative control inappropriate.

MPC uses a model of the plant for the prediction. Based on a specified cost criterion, the control inputs are computed such that this cost criterion is minimal for the predicted behavior. To ensure robustness with respect to a model plant mismatch and measurement noise a feedback is introduced by measuring the current state and recomputing the optimal control inputs repeatedly. MPC is well studied in the academic environment leading to important theoretical results including rigorous stability proofs. Furthermore, MPC is successfully applied in numerous industries ranging from, e.g., process industry to electrical power conversion and automotive industry. This success is based on the advantages of MPC. MPC can be applied to many plants including nonlinear models and multiple inputs. Moreover, hard input as well as state constraints can be easily incorporated. Furthermore, a time domain performance criterion is approximately minimized.

For MPC, typically the solution of an optimization problem in real time is required. Hence, a computationally tractable prediction is essential for the application of MPC. Moreover, the requirements for controlled systems are continuously increasing, which results in more complex models. For example, the required fast adaption of production targets in the chemical industry to the market demand necessitates more frequent load changes [Agar et al., 2017], which calls for models that represent a whole operation regime and not only the area around one particular steady state. Or the desired increase in energy efficiency of chemical processes to reduce the CO_2 footprint results in more couplings between the reactors, e.g., by using the waste heat of one reactor for another reaction [Agar et al., 2017]. Another example is the increasing share of renewable energies in the power grid, which leads to an increase in complexity of the controlled system. This increase in model complexity makes the solution of the optimization problem in real time computationally challenging.

One remedy described in numerous articles is the use of reduced models for the prediction. Model reduction algorithms for nonlinear systems are much less developed than for linear systems. Furthermore, the complexity of the functional expression of the ordinary differential equation (ODE) is limited by the model order for linear time invariant (LTI) models but not for nonlinear models (compare with [Rewieński, 2003, Section 2.3]). Hence, we cover methods that reduce the order and complexity of nonlinear models.

Due to the mismatch between the reduced model and the plant, constraint satisfaction and stability of the closed loop are not guaranteed any more. To recover the important guarantees of MPC, it is essential to take the error of the model reduction into account. An intermediate step towards an MPC approach with robustness against the model reduction error is a bound for the error between the original and the reduced model while simulating the reduced model. There are only few results for MPC using reduced models that provide guarantees for the closed loop with the plant even for LTI systems. Hence, a bound for the model reduction error and an MPC approach using reduced models are developed for LTI systems.

Altogether, in this thesis we focus on three research directions. First, model reduction procedures that result in models with a reduced order and reduced complexity. Second, bounds for the model reduction error that can be utilized during simulation. Third, MPC using reduced models with a guaranteed behavior of the closed loop with the original model. For the last research direction, the fields model reduction, error bounds, and MPC have to be combined as visualized in Figure 1.1.

A connection between all methods developed in this thesis are *reduced models* that are either computed by model reduction methods or used in MPC schemes. In addition to the connection in form of reduced models, the methods proposed in this thesis are motivated by two shared features, which are, inter alia, important for the applicability in industry: *computational efficiency* and rigorous *guarantees*. For the model reduction methods, the aspect of computational complexity appears twice: the procedure needs to be computationally tractable and the resulting models possess a low complexity. Guarantees for the model reduction are, e.g., preservation of stability. By simply using reduced models it is unclear how the approximation error deteriorates the application at hand. Hence, it is crucial to provide rigorous bounds for the error between the original and the reduced model. For these bounds, it is important to establish a good compromise between a low computational complexity and a small conservatism. For MPC using reduced models the common guarantees of nominal MPC such as constraint satisfaction, bounds for the cost functional value, and asymptotic stability should be recovered. The computational efficiency shows up for MPC by using reduced models, error bounds with a low computational burden, and the way these ingredients are combined in the optimization problem.

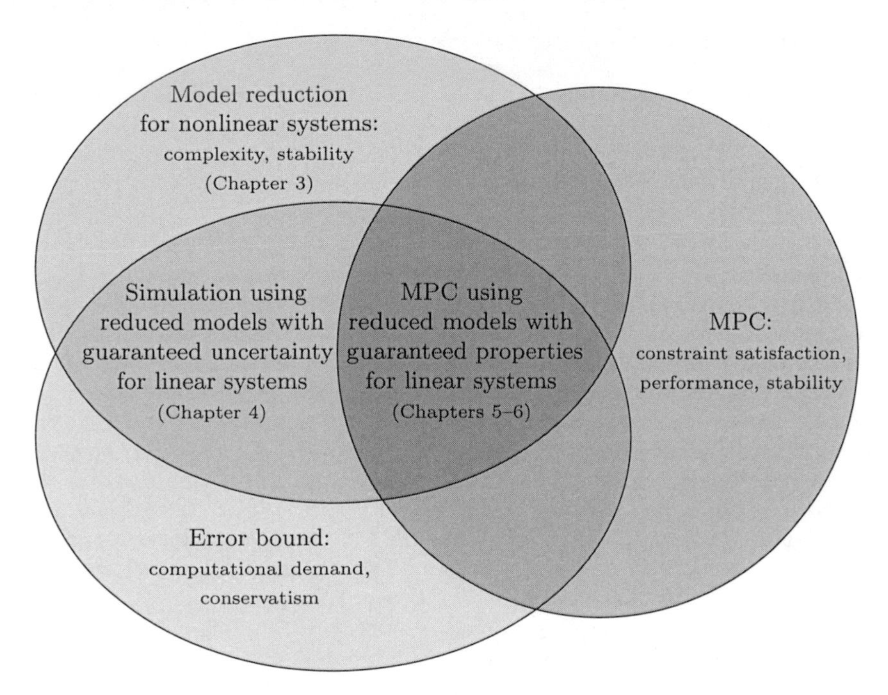

Figure 1.1: Research directions and aspects from each research field covered in this thesis.

1.2 Overview of the Research Area

In the last section, the three research areas of this thesis have been motivated:

i) model reduction

ii) error bounds for simulation using reduced models

iii) MPC using reduced models

In this section, we give a brief overview of related work in this three research areas and challenges that are addressed in this thesis. Hereby, we provide the basis for the contributions of this thesis, which are stated in Section 1.3.

An introduction into the three research areas is given in Chapter 2. If few prior knowledge of one research area exist, it is beneficial to read the corresponding section of Chapter 2 before this section.

Small parts of Section 1.2.2 and 1.2.3 have already been presented in [Löhning et al., 2014].

1.2.1 Model Reduction

In this section, we give a brief overview of the field of model reduction. For a more thorough introduction we refer for linear model reduction to [Antoulas, 2005a,b; Baur et al., 2014; Benner et al., 2015, 2017; Besselink et al., 2013; Obinata and Anderson, 2001] and for nonlinear model reduction to [Baur et al., 2014; Marquardt, 2002; Đukić and Sarić, 2012].

For linear systems, well-developed methods for model reduction are described in the literature.

The well-known method of balanced truncation introduced in [Moore, 1981; Mullis and Roberts, 1976] uses a state space realization in which each state is as well controllable as it is observable. Then the states that are simultaneously difficult to observe and difficult to control are truncated. A survey of model reduction by balanced truncation can be found in [Gugercin and Antoulas, 2004]. Under mild assumptions, balanced truncation preserves stability [Pernebo and Silverman, 1982] and an error bound [Enns, 1984; Glover, 1984] exists. Furthermore, it has been generalized to unstable [Kenney and Hewer, 1987; Therapos, 1989; Zhou et al., 1999] systems, passivity preservation [Desai and Pal, 1984], and frequency-weighted balanced truncation [Enns, 1984]. Closely related to balanced truncation is the Hankel norm approximation developed by Glover [1984].

Another established method is moment matching, which finds a reduced model that matches the first derivatives of the transfer function at a certain number of points. In the 1990s, numerically efficient algorithms have been developed for moment matching [Feldmann and Freund, 1995; Grimme, 1997; Odabasioglu et al., 1998]. Algorithms are available for preserving passivity [Odabasioglu et al., 1998]. Further details can be found, e.g., in [Benner et al., 2017, Chapter 7] and [Antoulas, 2005b; Freund, 2003].

A frequently used model reduction method is modal truncation [Antoulas, 2005a; Varga, 1995]. Modal truncation utilizes the eigenvalue decomposition of the state matrix and preserves the n_{R} dominant poles in the reduced model. Several measures for the dominance of poles exist, e.g., distance from the imaginary axis or the Hankel singular values of the subsystems corresponding to each block of the state matrix [Varga, 1995]. Another method is the proper orthogonal decomposition (POD) [Holmes et al., 1996; Lumley, 1967; Rathinam and Petzold, 2003; Sirovich, 1987], which uses state trajectories to compute a subspace on which the detailed model model is projected. For more details of the POD we refer to Section 2.1.2. Both, the modal truncation as well as the POD are applicable to high-dimensional dynamical system.

Optimization based approaches to model reduction connected to the POD are proposed in [Bui-Thanh et al., 2007; Kunisch and Volkwein, 2008]. Bui-Thanh et al. take the ODE of the reduced model into account in the computation of the subspace used for projection. Kunisch and Volkwein consider the case that the reduced model is used for optimal control and consider the dependence of the subspace used for projection on the control input.

The above methods for linear model reduction rely on a projection of the detailed model as discussed below in Section 2.1.1. If the internal dynamics are unknown and only input/output (I/O) data is available, e.g., a time-domain Loewner approach [Peherstorfer et al., 2017] or transfer function fitting [Sootla, 2013; Sou et al., 2008] can be used.

In contrast to model reduction for linear systems, model reduction for nonlinear systems is much less developed [Benner et al., 2015]. Many model reduction methods for nonlinear systems are extensions of methods for linear systems [Baur et al., 2014], e.g., balanced truncation [Fujimoto and Scherpen, 2010; Scherpen, 1993] and moment matching [Astolfi, 2010; Ionescu and Astolfi, 2016]. Unfortunately, both methods require the solution of partial differential equations (PDEs). Hence, they are computationally involved. A remedy for balanced truncation is the empirical balanced truncation [Kawano and Scherpen, 2017; Lall et al., 2002; Pallaske, 1987]. Empirical balanced truncation as well as the POD require trajectories generated by simulation and rely on a linear projection. But, according to Gu [2011], "nonlinear projection is natural and appropriate for reducing nonlinear systems, and can achieve more compact and accurate reduced models than linear projection". The extensions of balanced truncation [Fujimoto and Scherpen, 2010; Scherpen, 1993] and moment matching [Astolfi, 2010; Ionescu and Astolfi, 2016] to nonlinear systems can be interpreted as a nonlinear projection. But these methods are computationally challenging.

There exist several methods relying on a linear mapping between the states of the detailed and reduced model. The nonlinear Galerkin projection [Matthies and Meyer, 2003] assumes a (linear) subspace for the dominant states. The nonlinear Galerkin projection results in a differential algebraic equation (DAE). In contrast to the empirical balanced truncation and the POD performing a transformation followed by truncation, the nonlinear Galerkin projection is strongly connected to residualization, which also results in a DAE as discussed briefly in Section 2.1.1. The trajectory piecewise linear approximation [Rewieński and White, 2006] linearizes the nonlinear system around several points in the state space. Then, the nonlinear system is approximated with a weighted sum of the linear systems. The order reduction is done by a projection of this sum of the linear systems onto one (linear) subspace. Gu [2009] suggests a method that deduces an equivalent quadratic-linear DAE first, which is then reduced using a linear projection.

The above cited methods either utilize a linear mapping between states of the detailed and the reduced model or a nonlinear transformation, which is difficult to compute. Alternative approaches rely on data sampled from trajectories. Lohmann [1994] presented a model reduction procedure assuming that the dominant states are known and the nonlinearity is kept in the reduced model. The linear couplings between the state variables and the nonlinear functions are computed by minimizing the equation error. Another optimization based approach proposed in [Bond et al., 2010] allows to, first, utilize a reduced order training set of the states computed by a linear projection and, second, enforce incremental stability. The reduced basis methodology provides a framework for iteratively adding new trajectories to the training set (known as greedy sampling) and algorithms yielding a basis used for projection [Haasdonk and Ohlberger, 2008]. In [Wood et al., 2004], derivatives of I/O data up to the second-order are used to determine an implicit nonlinear ODE. Vargas and Allgöwer [2004] presented a procedure that is applicable to systems admitting a discrete-time Volterra representation and suggested an iterative approach for the construction of the reduced model. In this thesis, we build upon the work in [Lohmann, 1994; Vargas and Allgöwer, 2004; Wood et al., 2004].

Above, the aspects of a small approximation error and computational efficiency of the procedure are discussed. Another important issue of a model reduction procedure is preserving properties of the model such as stability or passivity [Antoulas, 2005b; Astolfi, 2010] and thereby providing guarantees for the model reduction procedure. For nonlinear systems, preserving local asymptotic stability of the equilibrium was shown for balanced truncation [Scherpen, 1993]. For moment matching conditions exist that ensure a locally asymptotically stable equilibrium for the reduced model [Astolfi, 2010; Ionescu and Astolfi, 2016]. Local asymptotic stability of the equilibrium is preserved for the POD if the detailed model is appropriately transformed in advance [Prajna, 2003]. For the trajectory piecewise linear approximation small-signal finite-gain $\mathcal{L}_p$ stability can be guaranteed [Bond and Daniel, 2009]. Model reduction preserving incremental stability is covered in [Besselink, 2012; Bond et al., 2010]. For incremental stable systems, state trajectories corresponding to a given input signal converge to each other [Angeli, 2002]. Hence, systems with multiple equilibria considered in Section 3.3 cannot be handled by incremental stability.

Another important issue is the computational complexity of the reduced model. For model reduction of linear systems the order of the model is often used as measure for model complexity. For general nonlinear systems, applying a linear projection reduces the order of the model but the complexity of the nonlinear expression remains similar. Hence, for nonlinear model reduction the complexity of the functional expressions is also important [Astolfi, 2010; Ionescu and Astolfi, 2016]. Simplification of the nonlinear expression before a linear projection can be achieved by a Taylor series expansion, which allows to compute the coefficients of the polynomials offline [Phillips, 2003]. Using the trajectory piecewise linear approximation [Rewieński and White, 2006] also results in a simplification of the nonlinear expression since it results in a weighted sum of linear systems. Furthermore, to reduce the complexity of the nonlinearity the empirical interpolation [Barrault et al., 2004; Chaturantabut and Sorensen, 2010; Drohmann et al., 2012b; Haasdonk et al., 2008; Peherstorfer et al., 2014] has been proposed.

1.2.2 Bounds for the Error of Model Reduction

Using a reduced model for simulation introduces uncertainty due to the mismatch between the detailed and reduced model. For the application of the reduced model it is important to quantify the uncertainty, especially in safety critical situations.

In the model reduction community several bounds for the model reduction error are known that are satisfied for all inputs. For linear systems, an a-priori error bound exists for balanced truncation with zero initial condition [Enns, 1984; Glover, 1984] and inhomogeneous initial condition [Baur et al., 2014; Heinkenschloss et al., 2011]. Error bounds for moment matching are proposed in [Panzer et al., 2013; Wolf et al., 2011]. In contrast to the a-priori error bounds, the a-posteriori error bounds are applicable after the reduced model has been computed. An a-posteriori error bound for stable (parameterized) linear ODEs is presented in [Haasdonk and Ohlberger, 2011]. In [Haasdonk and Ohlberger, 2011], the error is bounded by a scalar ODE, which depends only on the input and state of reduced model. Hence, this error bound takes the input trajectory into account. But the error bound monotonically increases with time. This is overcome by the generalized error bound introduced in [Ruiner et al., 2012]. For this error bound, the computational demand in the offline phase can be reduced significantly [Grunert et al., 2020]. Unfortunately, the error bound of [Grunert et al., 2020; Ruiner et al., 2012] requires the computation of a convolution integral for every time point since it cannot be written as an ODE. This significantly increases the computational demand. A-posteriori bounds for the error of the transfer function have been presented in [Antoulas et al., 2018; Feng et al., 2017]. These a-posteriori error bounds can be used to refine the reduced model [Antoulas et al., 2018].

A-posteriori error bounds exist also for nonlinear systems. For the trajectory piecewise linear approximation of stable systems an error bound is introduced in [Rewieński and White, 2003]. An error bound for the Galerkin projection with a POD basis is presented in [Volkwein, 2011]. An error bound for any projection based model reduction and approximation of the nonlinearity with the discrete-empirical interpolation method is introduced in [Wirtz et al., 2014]. Furthermore, an error bound for incremental balanced truncation exist [Besselink, 2012].

When reduced models are used for MPC as considered in the subsequent section, several error bounds have been utilized. Narciso and Pistikopoulos [2008] proposed to use the a-priori error bound of balanced truncation. In [Dubljevic et al., 2006], the model reduction method is limited to modal truncation, where the states of the reduced model and the neglected dynamics are only coupled by the input. Then, input-to-state boundedness of the error system is exploited to establish a constant error bound. In [Kögel and Findeisen, 2015; Lorenzetti et al., 2019; Sopasakis et al., 2013], the error bounds are based on robust positive invariant sets while no assumptions on the model reduction method are stated. Kögel and Findeisen [2015] compute, in a first step, box constraints for the error of the performance output and the error in the dynamics of the estimated states of the reduced model. These box constraints bound these errors for all possible trajectories in a chosen time interval. In a second step, these box constraints are utilized to compute a robust positive invariant set for the estimated state of the reduced model. Finally, the error bounds on the input and performance output are established using the box constraints and the robust positive invariant set. In [Lorenzetti et al., 2019] the error bounds are directly computed based on the given dynamics. This results in less conservative error bounds. In the recent publication [Lorenzetti and Pavone, 2019], an error bound is proposed that bounds the contribution from the most recent time interval by polytopic sets and all prior contributions by the a-posteriori error bound presented in our work [Löhning et al., 2014]. To compute the polytopic set, all inputs and states in a bounded set are taken into account. To allow for the a-posteriori error bound, a projection-based model reduction method is assumed in [Lorenzetti and Pavone, 2019]. Altogether, in [Dubljevic et al., 2006; Kögel and Findeisen, 2015; Lorenzetti and Pavone, 2019; Lorenzetti et al., 2019; Narciso and Pistikopoulos, 2008; Sopasakis et al., 2013], only error bounds that are fulfilled for all inputs or all inputs and states in a bounded set have been utilized for MPC using reduced models. Hence, the known input and state of the reduced model is not taken into account in the prediction of the error bound, which, in general, results in a considerable conservatism.

1.2.3 MPC Using Reduced Models

In this section, we discuss the literature in the field of MPC related to this thesis. At the beginning, we discuss the literature ensuring stability of the closed-loop system. Afterwards, we concentrate on MPC using reduced models. For a general overview of MPC we refer to the books [Camacho and Bordons, 2007; Grüne and Pannek, 2017; Kouvaritakis and Cannon, 2016; Rawlings and Mayne, 2009] and survey articles [Findeisen et al., 2003; Joe Qin and Badgwell, 2003; Magni and Scattolini, 2004; Mayne, 2014; Mayne et al., 2000].

In MPC, the infinite horizon optimal control problem is approximated by a finite horizon optimal control problem. Hence, stability of the closed-loop system is not guaranteed a-priori. Many versions of MPC have been proposed in the literature in order to ensure stability. A zero terminal state constraint was used in [Chen and Shaw, 1982; Keerthi and Gilbert, 1988; Mayne and Michalska, 1990]. The zero terminal state constraint was extended to a terminal constraint set [Chisci et al., 1996; Michalska and Mayne, 1993; Scokaert et al., 1999]. This MPC version is known as dual mode since inside the terminal constraint set a local control law is used instead of the model predictive controller. A terminal cost (without terminal state constraint) was proposed for linear, constrained, and stable systems in [Rawlings and Muske, 1993]. The combination of a terminal constraint set and terminal cost emerged to a common framework to guarantee stability for MPC [Chen, 1997; Chen and Allgöwer, 1998; Chmielewski and Manousiouthakis, 1996; De Nicolao et al., 1998; Fontes, 2001; Scokaert and Rawlings, 1998]. The terminal constraint set is often defined by a local stabilizing controller. In contrast to dual mode MPC, this local controller is never applied. But, the local stabilizing controller is used to prove recursive feasibility, i.e., from feasibility of the finite horizon optimal control problem at the initial time instant follows feasibility for all subsequent sampling instants.

Together with results from tube-based robust MPC [Bemporad and Morari, 1999; Mayne et al., 2005], the above references are the basis for the results about

MPC using reduced models of this thesis. Further results about stability of MPC are, e.g., contractive MPC [de Oliveira Kothare and Morari, 2000; Polak and Yang, 1993a,b; Yang and Polak, 1993] and unconstrained MPC [Grimm et al., 2005; Grüne, 2009; Grüne and Pannek, 2017; Grüne and Rantzer, 2008; Jadbabaie and Hauser, 2005; Reble, 2013; Reble and Allgöwer, 2012]. Unconstrained MPC relies on certain controllability assumptions to compute a sufficiently large prediction horizon in order to guarantee stability without a terminal constraint.

When MPC is applied to high-dimensional systems, solving the high-dimensional optimization problem is a large computational burden. Therefore, reduced models are frequently used for the prediction in MPC, e.g., [Agudelo et al., 2007a; Balasubramhanya and Doyle III, 2000; Dubljevic et al., 2006; Froisy, 2006; Hovland and Gravdahl, 2008; Hovland et al., 2008a; Huisman and Weiland, 2003; Jarmolowitz et al., 2009; Marquez et al., 2013; Nagy et al., 2000; Narciso and Pistikopoulos, 2008; Ou and Schuster, 2009; Shang et al., 2007; Touretzky and Baldea, 2014; Xie and Theodoropoulos, 2010]. Furthermore, linear reduced models can be exploited in MPC to facilitate the solution of the online optimization problem [Huisman and Weiland, 2003; Marquez et al., 2013]. For explicit MPC either reduced models or a projection of the state may be used to reduce the number of regions [Hovland et al., 2008a; Johansen, 2003]. Alternatively to using reduced models for MPC, the computational efficiency can also be increased by reformulating the optimization problem. Examples are move-blocking [Cagienard et al., 2007], generalized input parameterizations [van Donkelaar et al., 1999], or the approximation of the optimization problem [Kouvaritakis et al., 2002].

By using a reduced model within the model predictive controller, a mismatch between the plant and the prediction model is introduced. As a result, satisfaction of constraints or asymptotic stability of the closed loop may be lost. Since these are important properties of the closed-loop system, robustness against the model reduction error has to be ensured by the MPC scheme.

Narciso and Pistikopoulos [2008] proposed to utilize the a-priori error bound of balanced truncation to tighten the output constraints. However, recursive feasibility and asymptotic stability are not established in [Narciso and Pistikopoulos, 2008]. When the high-dimensional system is given by a PDE, under the assumption of recursive feasibility, asymptotic stability of the closed-loop system and hard state constraint satisfaction can be guaranteed [Dubljevic et al., 2006]. However, in [Dubljevic et al., 2006], the model reduction method is limited to modal truncation.

Alternatively, one may think of methods from robust MPC [Bemporad and Morari, 1999; Mayne et al., 2005]. As noted in [Hovland et al., 2008a], "the applicability of these methods to establish robustness in the context of MPC with reduced-order models remains a challenging open research question". In the meantime, results of robust output feedback MPC [Løvaas et al., 2007, 2008a,b] have been specialized to MPC using reduced models. Hovland et al. [2008b] guarantee robust stability despite the model reduction error by choosing the cost functions such that a Lyapunov function for the closed-loop system decreases with time. The work of Hovland et al. does not rely on an explicit bound on the model reduction error with the result that it applies only to stable systems and furthermore soft state constraints. In [Sopasakis et al., 2013], it is suggested to use methods from tube-based robust MPC [Mayne et al., 2005, 2006] to establish constraint satisfaction and asymptotic stability of a (possibly large) set around the origin. Tube-based robust MPC is exploited in [Kögel and Findeisen, 2015; Lorenzetti and Pavone, 2019; Lorenzetti et al., 2019] to prove constraint satisfaction and asymptotic stability of a (possibly large) set around the origin despite the model reduction error and a bounded additive disturbances on the system dynamics and measurement. Furthermore, robust MPC is utilized in [Bäthge et al., 2016] to show recursive feasibility when using a coarse or reduced model for the long-term prediction. The error from using the coarse model is approximated by an additive uncertainty and an uncertain initial condition. Hence, constraint satisfaction and recursive feasibility is not proven when a reduced model is used for the long-term prediction.

Altogether, asymptotic stability and satisfaction of hard state constraints for MPC using a reduced model is treated, to the best of our knowledge, only in [Dubljevic et al., 2006; Kögel and Findeisen, 2015; Lorenzetti and Pavone, 2019; Lorenzetti et al., 2019; Sopasakis et al., 2013]. However, in all these references, satisfaction of hard state or output constraints is guaranteed by tightening the constraints according to a bound on the error between the detailed and the reduced model. To predict this bound, all inputs or all inputs and states in a bounded set are taken into account. Thereby, significant conservatism is introduced, since this possibly large error bounding sets are used to tighten the constraints.

The existing methods for MPC using reduced models with guarantees are compared in Table 1.1 with respect to the applicable model reduction method, conservatism of the error bound utilized to satisfy hard state constraints, asymptotic stability of the origin, computational efficiency of the optimal control problem, and the possibility to use output feedback. The method proposed in this thesis is also shown in Table 1.1 to ease the comparison with existing methods.

1.2.4 Summary

In the last sections, we have given an overview about the literature concerning model reduction, bounds for the reduction error, and MPC using reduced models.

While for model reduction of linear systems several widely accepted methods that result in models with reduced order and reduced computational complexity exist, methods for nonlinear systems are less developed. Challenges in nonlinear model reduction are, first, that a nonlinear mapping between the states of the detailed and the reduced model can be required to achieve a reduced model of low order despite a small approximation error. Second, for nonlinear systems besides a reduced order also a simplification of the functional expression is often necessary to achieve computationally efficient reduced models. Third, system properties such

Table 1.1: Properties of methods for MPC using reduced models with robustness against the model reduction error. A ✓ (–, ×) denotes that the property is satisfied largely (with restrictions, not at all or with considerable restrictions).

	Model reduction	Error bound and constraint satisfaction	Asymptotic stability of origin	Computational efficiency	Output feedback
Modal truncation, error bound for worst-case input, recursive feasibility assumed [Dubljevic et al., 2006]	×	–	–	✓	×
Balanced truncation, a-priori error bound of balanced truncation [Narciso and Pistikopoulos, 2008]	–	–	×	✓	×
Any model reduction method, no explicit error bound, asymptotic stability of the origin by constraints on the cost function [Hovland et al., 2008b]	✓	×	✓	✓	✓
Any model reduction method, error bound for worst-case inputs, asymptotic stability of a set around the origin [Sopasakis et al., 2013]	✓	–	–	✓	×
Any projection-based model reduction method, error bounding system taking the actual input and state into account, asymptotic stability of the origin [Löhning et al., 2014]	✓	✓	✓	✓	×
Any model reduction method, error from model reduction approximated by additive bounded uncertainty and uncertain initial condition [Bäthge et al., 2016]	✓	×	×	✓	×
Any (projection-based) model reduction method, error bound for all inputs and states in a bounded set, asymptotic stability of a set around the origin [Kögel and Findeisen, 2015; Lorenzetti and Pavone, 2019; Lorenzetti et al., 2019]	✓	–	–	✓	✓

as stability need to be preserved in many cases. Fourth, it is desired to obtain a computationally efficient model reduction procedure.

For the application of reduced models we have looked at MPC using reduced models. For this field, the main challenge is to provide guarantees for the closed loop with the plant such as asymptotic stability of the origin and satisfaction of constraints with a reasonable conservatism while allowing for a computationally efficient solution of the online optimization problem. Even for linear systems only few and limited results exist. To tackle the challenge, bounds for the model reduction error are required that are both tight and computationally efficient. Moreover, existing applicable a-posteriori error bounds are only marginally stable, which prevents to establish asymptotic stability of the origin for the closed loop with the plant. Hence, another challenge is to derive error bounds that converge to zero asymptotically for vanishing input.

1.3 Contributions of the Thesis

In this thesis, we address the challenges described in the previous section. Thereby, we contribute to the fields of model reduction and MPC. More precisely, we present novel methods for the research topics model reduction of nonlinear systems, bounds for the model reduction error, and MPC using reduced models.

With respect to model reduction, we propose in Chapter 3

- a model reduction method for nonlinear continuous-time dynamical systems, which allows to obtain models of low order and low computational complexity.

The method is an I/O trajectory-based approach that — in contrast to many other existing model reduction methods — relies on parameter optimization and not on a projection. This allows us to use a nonlinear mapping from the state variables of the detailed model to the state variables of the reduced model determined by the observability map. A low complexity functional expression of the reduced model is achieved by sparsity enhancing ℓ_1-minimization. Moreover,

- we extend the proposed method to preserve the location and local exponential stability of multiple steady states.

For this purpose, we derive a necessary and sufficient condition for the simultaneous stability of a set of steady states. We relax the resulting optimization problem to a sequential convex optimization problem, which admits an efficient optimization. The reduced model with low complexity obtained using the proposed method can be used for MPC of complex nonlinear systems. MPC using reduced models with guaranteed asymptotic stability and constraint satisfaction is demanding even for LTI systems as indicated by the few and limited results. To overcome the limitations depicted in Table 1.1, we focus on the class of LTI systems for the error bound and the MPC schemes.

In Chapter 4, we improve an existing a-posteriori bound for the model reduction error. This results in

- an asymptotically stable system that bounds the error between the detailed and the reduced model.

Due to the asymptotic stability instead of marginal stability, the improved error bound is considerably tighter than the original one while at the same time possesses a comparable computational demand. For the application in model based control, we achieve the asymptotic stability even for unstable systems by prestabilization. Furthermore, we compare the proposed error bound with existing ones using a model of a tubular reactor.

In Chapter 5, we utilize the improved error bound to derive

- an MPC scheme using a reduced model that guarantees asymptotic stability and satisfaction of hard input and state constraints

when the model predictive controller is applied to a high-dimensional and possibly unstable plant. Besides asymptotic stability and constraint satisfaction, obtaining a good performance with respect to the cost functional is an important advantage of MPC. Hence, we show that for a common choice of design parameters, first, the infinite horizon cost functional is preserved while replacing the plant with the reduced model and, second,

- the proposed MPC scheme implicitly minimizes the quasi-infinite horizon cost functional of an MPC scheme using the plant model despite using a reduced model for the prediction.

In this case, the proposed MPC scheme also guarantees an upper bound for the cost functional value of the closed loop with the plant. To achieve the mentioned guarantees, the online optimization problem includes the nonlinear error bounding system. This raises the question of the computational complexity of the online optimization problem. Therefore, in Chapter 6 we show for discrete-time plants that

- the online optimization problem can be reformulated as a second-order cone program,

which can be solved efficiently. Existing MPC schemes using reduced models with comparable guarantees have been applied to a practically motivated example, to the best of our knowledge, only in [Kögel and Findeisen, 2015]. We demonstrate the applicability of the developed MPC scheme to control the model of a tubular reactor. In contrast to [Kögel and Findeisen, 2015], we show in the simulation study that

- the proposed MPC scheme achieves a good trade-off between computational efficiency and conservatism

while at the same time providing important guarantees for the closed-loop behavior.

Summarizing, we provide novel methods concerning model reduction and the utilization of reduced models with the common goal of reduced computational complexity. Since the model reduction error can compromise the application at hand, we provide methods with rigorous guarantees in this thesis. Our main contributions are a novel method for model reduction of nonlinear systems, improved a-posteriori bounds for the model reduction error as well as novel algorithms for the utilization of reduced models in MPC.

1.4 Design Workflow of the Proposed Model Predictive Control Scheme

One major contribution of this thesis is the proposed MPC scheme using a reduced model and error bound. This MPC scheme combines several elements introduced in different chapters. As a consequence, the workflow to design the proposed model predictive controller using a reduced model depicted in Figure 1.2 serves the reader also as a guide while reading the thesis. Hence, we present the workflow already here.

The first step to design the proposed model predictive controller using a reduced model is preprocessing of the plant model in order to end up with an asymptotically stable preprocessed model. Then, the reduced model is computed by projection of the preprocessed model. In this thesis, the POD is used for this step but any projection based methods are applicable. In the third step, the error bound is derived and evaluated. If the error bound is too conservative, the preprocessing or the model reduction has to be adapted. If the error bound is tight enough, the proposed model predictive controller can be designed. To achieve a satisfactory closed-loop performance, further adaptions of the design parameters of all steps can be required.

1.5 Outline of the Thesis

The background for this thesis is provided in Chapter 2. This includes the framework of model reduction by projection, model reduction by POD, existing a-posteriori bounds for the model reduction error, a common framework to guarantee stability for nominal MPC, and the control problem of a tubular reactor.

In Chapter 3, we present a method for model reduction of nonlinear systems based on input-output data. After the problem statement, the procedure is described and utilized to reduce the model of a mitogen-activated protein kinase (MAPK) cascade. In the third section of Chapter 3, the procedure is extended to preserve the location and local exponential stability of multiple steady states.

The model reduction introduces an error between the detailed model and the reduced model. Hence, in Chapter 4 we show for LTI systems how this error can be bounded while simulating the reduced model. Although the error bound can be used in a broad context, we aim at the application for MPC. After stating the problem setup we introduce a preprocessing of the plant including a prestabilization and a state transformation. Then, a bound for the norm of the matrix exponential

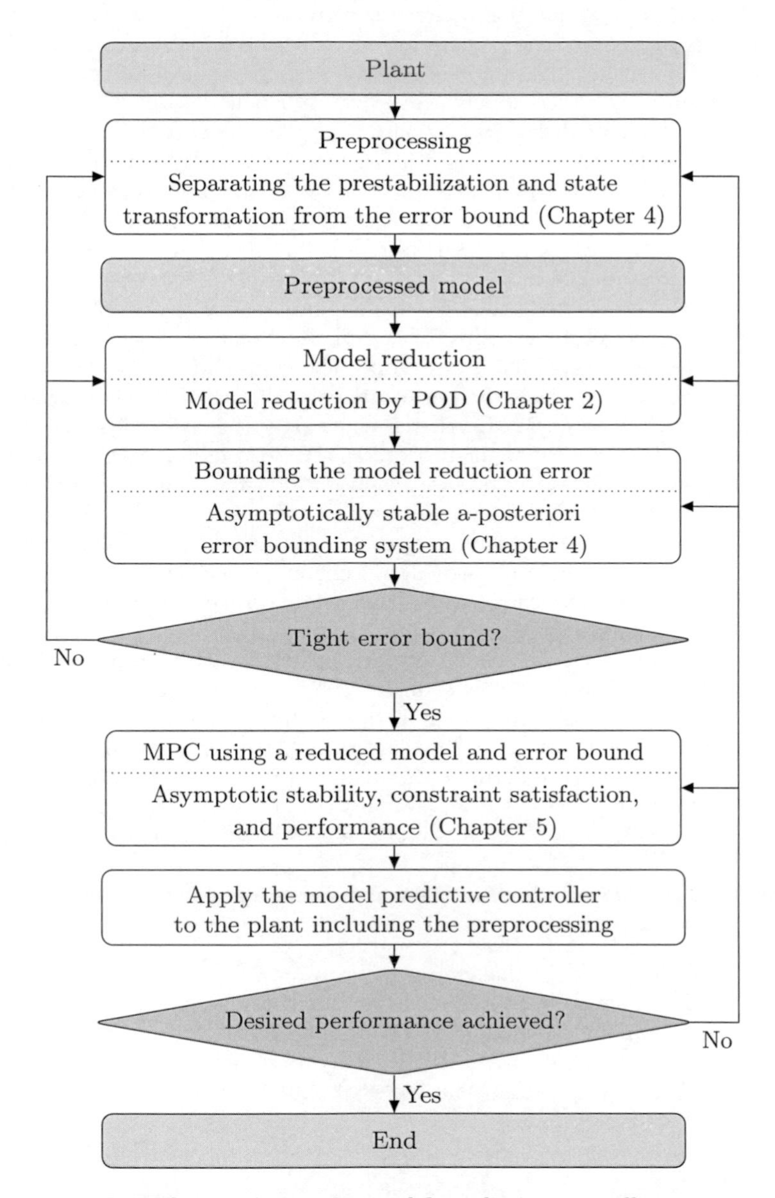

Figure 1.2: Workflow to design the model predictive controller using a reduced model and error bound.

is introduced and the connection between the preprocessing and the parameters of this bound is discussed. The a-posteriori error bound is presented in the fourth section together with a discussion of several ways to achieve an asymptotically stable error bound and the relation to existing error bounds. Finally, the proposed error bound is compared with existing ones using the model of the tubular reactor.

In Chapter 5, the error bound is utilized for MPC using reduced models. At the beginning we state the problem setup and apply the preprocessing and model reduction. Then, we ensure constraint satisfaction for the preprocessed model in a computationally efficient way. Furthermore, we show that for a certain choice of design parameters the model reduction error can be eliminated in the cost functional. The main theoretical result of Chapter 5 is the proof of recursive feasibility and asymptotic stability of the plant model in closed loop with the proposed MPC scheme. Finally, we apply the proposed MPC scheme to the tubular reactor and compare the performance with an MPC scheme using the plant model as well as an MPC scheme using only the reduced model without the error bound. Using the plant model for MPC is possible in this simulation study but would in general be prohibitive in an industrial application due to the real-time requirements.

The application of MPC requires optimization in real time. Thus, the computational demand of the proposed MPC scheme is considered in Chapter 6 in order to enhance the applicability. To have a finite-dimensional static optimization problem, we consider discrete-time systems. The main goal is to deduce a computationally efficient formulation for the optimization problem. Thus, we show that the optimization problem can be reformulated as a convex optimization problem. Afterwards, the convex optimization problem is reformulated such that it is independent of the dimension of the plant model. Furthermore, the optimization problem is stated as a second-order cone program (SOCP), which allows to utilize dedicated and more efficient solvers. Finally, the computational demand of the three MPC schemes considered in Chapter 5 is assessed by means of the tubular reactor.

This thesis concludes with a summary and discussion followed by an outlook of possible future research directions.

Some results of this thesis have already been published in a very similar form. Parts of Chapter 2 have already been presented in [Löhning et al., 2014]. Chapter 3 is very similar to [Löhning et al., 2011a,b]. Chapter 4 is partially based on [Hasenauer et al., 2012; Löhning et al., 2011c, 2014]. Preliminary results of Chapter 5 have already been published in [Löhning et al., 2014].

Chapter 2

Background

In the previous chapter, we have given an overview of the research areas related to this thesis and we have outlined the contributions of this thesis. In this chapter, we describe existing results underlying the contributions in more detail to provide the knowledge required subsequently. In Section 2.1, we present a common framework for model reduction and show one particular method for model reduction known as POD. For this model reduction framework, we discuss in Section 2.2 existing a-posteriori error bounds. An introduction to MPC and a common framework to guarantee convergence to the origin for nominal MPC is given in Section 2.3. Furthermore, we introduce an MPC approach using a reduced model that neglects the model reduction error. Finally, in Section 2.4, we introduce the model of a nonisothermal tubular reactor, which is used in Chapters 4–6 to evaluate the presented methods.

Parts of Sections 2.3 and 2.4 have already been presented in [Löhning et al., 2014].

2.1 Model Reduction

The field of model reduction deals with algorithms that simplify dynamical models. This system to be reduced will be called detailed model in this thesis and abbreviated using the subscript D. We consider detailed models described by n first-order ODEs and a set of n_{Y} algebraic equations defining the output $y_{\mathrm{D}}(t; x_0, u) \in \mathbb{R}^{n_{\mathrm{Y}}}$ at time $t \in \mathbb{R}_{0+}$ for the initial condition $x_0 \in \mathbb{R}^n$ and input $u(\cdot) \in \mathcal{L}_2^{n_{\mathrm{U}}}$ of dimension n_{U} as in [Antoulas, 2005b]. Hence, the detailed models are of the form

$$\Sigma_{\mathrm{D}} : \begin{cases} \dot{x}(t) = f\big(x(t), u(t)\big)\,, & x(0) = x_0\,, \\ y_{\mathrm{D}}(t) = h\big(x(t), u(t)\big)\,, \end{cases} \tag{2.1}$$

in which $x(t) \in \mathbb{R}^n$ is the state of the system at time $t \in \mathbb{R}_{0+}$ and n the order of the model. To ensure existence and uniqueness of solutions the vector field $f : \mathbb{R}^n \times \mathbb{R}^{n_{\mathrm{U}}} \to \mathbb{R}^n$ is assumed to be globally Lipschitz continuous.

Given a detailed model Σ_{D}, the objective is to find a model of reduced complexity Σ_{R} that provides a good approximation of the I/O behavior of Σ_{D}. In this thesis, the complexity of a system contains the number of states as well as the computational complexity of the right-hand side of the dynamics. The approximation quality can be measured, for example, by the norm of the output error, i.e., the difference in the output for the same input and corresponding initial condition. While for LTI models a good approximation for all inputs is often considered, this choice may lead to unnecessarily complex reduced models for general nonlinear systems. Hence, we consider the output error only for a finite time interval $[0, T_{\text{end}}]$ as well as important initial conditions and input trajectories. The important initial conditions and inputs can be specified by a subset of the state space and input space or importance weights as used in Chapter 3.

2.1.1 Model Reduction by Projection

In this section, we discuss a very common framework for reducing detailed models of the form (2.1), which we denote model reduction by projection. This framework is used by many well-known methods, among other things, balanced truncation, POD, modal truncation, and Krylov methods. Model reduction by POD will be explained in Section 2.1.2. For the other model reduction procedures we refer to [Antoulas, 2005b]. In addition to the popularity, model reduction by projection is important in this thesis since the error bounds discussed in Section 2.2 and Chapter 4 allow for any model reduction method relying on projection. Since the error bounds are derived for linear systems, only linear transformations are considered in this section.

For model reduction by projection, we consider the full-column rank matrices $V \in \mathbb{R}^{n\times n_{\mathrm{R}}}$ and $W \in \mathbb{R}^{n\times n_{\mathrm{R}}}$ satisfying $W^{\mathsf{T}}V = I_{n_{\mathrm{R}}}$. We derive the reduced model of order n_{R} similar to [Antoulas, 2005b, Section 1.1.1]. Consider the nonsingular matrix $T = \begin{bmatrix} V & \tilde{V} \end{bmatrix} \in \mathbb{R}^{n\times n}$ with $T^{-1} = \begin{bmatrix} W & \tilde{W} \end{bmatrix}^{\mathsf{T}}$ and the state transformation $x(t) := Tz(t)$. Then, we partition

$$z(t) =: \begin{bmatrix} z_1(t) \\ z_2(t) \end{bmatrix} = \begin{bmatrix} W^{\mathsf{T}}x(t) \\ \tilde{W}^{\mathsf{T}}x(t) \end{bmatrix}$$

into the dominant states $z_1(t) \in \mathbb{R}^{n_{\mathrm{R}}}$ and the nondominant states $z_2(t) \in \mathbb{R}^{n-n_{\mathrm{R}}}$. By inserting $x(t) = Tz(t)$ into (2.1), we get

$$\begin{aligned} \dot{z}_1(t) &= W^{\mathsf{T}}f\big(Vz_1(t) + \tilde{V}z_2(t), u(t)\big)\,, \qquad & z_1(0) &= W^{\mathsf{T}}x_0\,, \\ \dot{z}_2(t) &= \tilde{W}^{\mathsf{T}}f\big(Vz_1(t) + \tilde{V}z_2(t), u(t)\big)\,, \qquad & z_2(0) &= \tilde{W}^{\mathsf{T}}x_0\,. \end{aligned}$$

The model reduction occurs by neglecting the dynamics of the nondominant states $z_2(t)$. Truncating the nondominant states results in the reduced model

$$\Sigma_{\mathrm{R}} : \begin{cases} \dot{x}_{\mathrm{R}}(t) = W^{\mathsf{T}}f\big(Vx_{\mathrm{R}}(t), u(t)\big)\,, \qquad x_{\mathrm{R}}(0) = z_1(0) = W^{\mathsf{T}}x_0\,, \\ y_{\mathrm{R}}(t) = h\big(Vx_{\mathrm{R}}(t), u(t)\big)\,. \end{cases} \tag{2.2}$$

An estimate for the state of the detailed model $x(t) = Vz_1(t) + \tilde{V}z_2(t)$ is given by $Vx_{\mathrm{R}}(t)$. Since the summand $\tilde{V}z_2(t)$ is neglected in the reduced model, the state of the reduced model $x_{\mathrm{R}}(t)$ and $z_1(t) = W^{\mathsf{T}}x(t)$ are, in general, not equal for $t > 0$.

In the last paragraph, we have seen that model reduction by projection is a truncation of states in an appropriate basis. If only the matrices V and W are given and not the whole transformation matrix T, as in model reduction by POD described below, an alternative derivation of the same reduced model Σ_R can be used. This alternative uses the projection VW^T onto the subspace spanned by the columns of V parallel to the kernel of W^T. In this approach, the state of the detailed model $x(t)$ is replaced in (2.1) by the estimated state $Vx_\mathrm{R}(t)$. Then, the resulting dynamics $V\dot{x}_\mathrm{R}(t) = f\big(Vx_\mathrm{R}(t), u(t)\big)$ are projected along the kernel of W^T. This results in the same reduced model Σ_R since $W^\mathsf{T}V = I_{n_\mathrm{R}}$.

Model reduction by projection is visualized in Figure 2.1. The initial condition of the detailed model is projected along the kernel of W^T onto the subspace spanned by the column of V. The reduced model evolves in the subspace spanned by the column of V. The black lines depict the connection between the state of the detailed and reduced model at the time instants $0, 0.5, 1, \ldots$. Clearly, the majority of these lines are not parallel to the kernel of W^T. This demonstrates that the state of the reduced model is, in general, not equal to the projected state of the detailed model for $t > 0$.

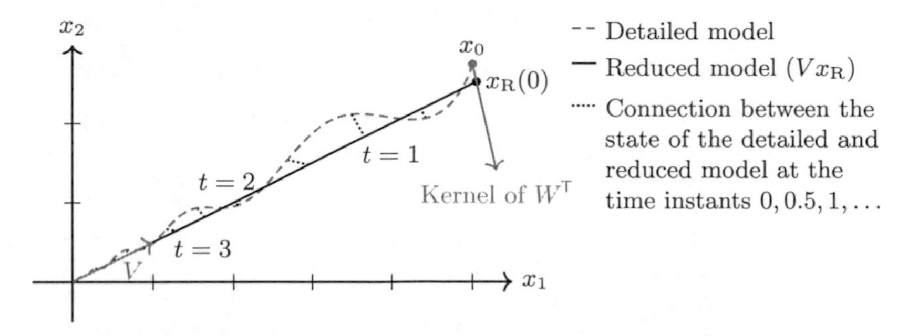

Figure 2.1: Visualization of model reduction by projection for $V = [1\ 0.5]^\mathsf{T}$ and $W = [0.9\ 0.2]^\mathsf{T}$.

Beyond truncation, the nondominant states can also be determined by an algebraic equation, as discussed in a similar context in [Marquardt, 2002]. Thereby, a connection to methods mentioned in Section 1.2 is obtained. One example is residualization, which assumes $\dot{z}_2(t) = 0 = W^\mathsf{T} f\big(Vz_1(t) + Vz_2(t), u(t)\big)$. Another example is the nonlinear Galerkin projection [Matthies and Meyer, 2003], which uses an explicit relation $z_2 = \eta(z_1)$ in which η is a vector field of appropriate dimension. These methods are not considered as model reduction by projection in this thesis as in [Antoulas, 2005b].

To evaluate the right-hand side of the dynamics of the reduced model (2.2), we have to compute, first, the n-dimensional estimated state $Vx_\mathrm{R}(t)$, then the vector field $f(\cdot)$, and finally the multiplication with W^T. Hence, in general, without any simplification of the functional expression $W^\mathsf{T} f(V\cdot)$, the computational complexity of the right-hand side remains constant for model reduction by projection [Chaturantabut, 2011, Section 2.1.2]. A notably exception are LTI systems

$$\Sigma_{\mathrm{D}}: \begin{cases} \dot{x}(t) = Ax(t) + Bu(t)\,, & x(0) = x_0\,, \\ y(t) = Cx(t) + Du(t)\,, \end{cases} \tag{2.3}$$

which result in the reduced model

$$\Sigma_{\mathrm{R}}: \begin{cases} \dot{x}_{\mathrm{R}}(t) = A_{\mathrm{R}}x_{\mathrm{R}}(t) + B_{\mathrm{R}}u(t)\,, & x_{\mathrm{R}}(0) = W^{\mathsf{T}}x_0\,, \\ y_{\mathrm{R}}(t) = C_{\mathrm{R}}x_{\mathrm{R}}(t) + D_{\mathrm{R}}u(t)\,, \end{cases} \tag{2.4}$$

in which $A_{\mathrm{R}} = W^{\mathsf{T}}AV$, $B_{\mathrm{R}} = W^{\mathsf{T}}B$, $C_{\mathrm{R}} = CV$, and $D_{\mathrm{R}} = D$. For the LTI case, the expressions $W^{\mathsf{T}}AV$, $W^{\mathsf{T}}B$, and CV can be computed beforehand. Therefore, a simplification of the functional expression is less important for LTI systems.

In the presented projection framework, the reduced model is uniquely determined by the detailed model and the matrices V, W. Hence, the problem of model reduction is to find these two matrices in an appropriate way. In the following subsection, we present one possible method, which will be used several times within this thesis.

2.1.2 Model Reduction by Proper Orthogonal Decomposition

A commonly used model reduction method is the proper orthogonal decomposition (POD). According to Antoulas [2005b], POD was the only systematic and widely used method for model reduction of nonlinear systems. In this thesis, model reduction based on POD is used in the examples of Chapters 4–6.

POD, which is also known as Karhunen-Loève decomposition and principal component analysis, has been used in the area of computational fluid dynamics to extract coherent structures [Lumley, 1967] and, later, for model reduction of dynamical systems [Sirovich, 1987]. Although POD can be used for infinite-dimensional models, e.g., for PDEs, we will follow the presentation in [Antoulas, 2005b] for finite-dimensional models. For a detailed introduction to POD we refer to [Holmes et al., 1996; Rathinam and Petzold, 2003; Volkwein, 2011] and [Schilders et al., 2008, page 95–109].

The main idea behind POD is that the time response $\{x(\cdot; x_0^{(i)}, u^{(i)})\}_{i=1}^{n_{\mathrm{T}}}$ to the relevant initial conditions $\{x_0^{(i)}\}_{i=1}^{n_{\mathrm{T}}}$ and input trajectories $\{u^{(i)}(\cdot)\}_{i=1}^{n_{\mathrm{T}}}$ contains the important behavior of the detailed model. Hence, the time response is used to compute a subspace on which the detailed model is projected onto as discussed in Section 2.1.1.

From the time responses we take n_{S} snapshots $x(t; x_0^{(i)}, u^{(i)})$ for every $i = 1, \ldots, n_{\mathrm{T}}$ at the time points $t = t_1, \ldots, t_{n_{\mathrm{S}}}$. The $n_{\mathrm{T}}n_{\mathrm{S}}$ snapshots are collected into the matrix $X \in \mathbb{R}^{n \times n_{\mathrm{T}}n_{\mathrm{S}}}$. We are looking for a set of $n_{\mathrm{R}} < \min(n, n_{\mathrm{T}}n_{\mathrm{S}})$ orthonormal vectors $v_i \in \mathbb{R}^n$, $i = 1, \ldots, n_{\mathrm{R}}$ such that an approximation of the snapshot matrix X in the subspace spanned by the vectors v_i minimizes the 2-induced norm of the error. A solution of the stated problem can be computed with the singular value decomposition (SVD) of $X = \check{U}\check{\Sigma}\check{V}^\mathsf{T}$. The first n_R columns of $\check{U}$ corresponding to the largest singular values are the requested orthonormal vectors v_i. Finally, the reduced model is computed by projection onto the span of the vectors v_i according to Section 2.1.1 with $V = W = \begin{bmatrix} v_1 & \dots & v_{n_\mathrm{R}} \end{bmatrix}$.

2.2 A-Posteriori Bounds for the Model Reduction Error

In the last section, we have seen that model reduction by projection results, in general, in an error in the truncated nondominant states as well as the dominant states. To bound these errors when simulating the reduced model, Haasdonk and Ohlberger [2011] proposed an a-posteriori error bound for linear time varying system with an affine parameter dependence of the system matrices. In this section, we introduce the error bound proposed in [Haasdonk and Ohlberger, 2011] for the subclass of LTI systems (2.3). This error bound is the basis for the error bound proposed in Chapter 4.

The error bound applies to any model reduction method relying on projection. Hence, we assume that the matrices V and W as introduced in Section 2.1.1 are given and result in the reduced model (2.4). For this setup, the following theorem adapted from [Haasdonk and Ohlberger, 2011] is applicable.

Theorem 2.1 (A-posteriori error bound)**.** *Consider the detailed model* (2.3) *and the reduced model* (2.4) *defined by* V, W*. If* $\alpha \geq 1$ *satisfies*

$$\left\| \mathrm{e}^{At} \right\| \leq \alpha \quad \text{for all } t \geq 0\,,$$

then, the error bound

$$\|x(t) - Vx_\mathrm{R}(t)\| \leq \Delta(t) := \alpha \big\| \big(I_n - VW^\mathsf{T}\big)x(0)\big\| + \alpha \int_0^t \|r(\tau)\|\, \mathrm{d}\tau \tag{2.5}$$

with the residual $r(t) := \big(I_n - VW^\mathsf{T}\big)\big(AVx_\mathrm{R}(t) + Bu(t)\big)$ *is satisfied for all* $t \geq 0$.

Proof. The proof follows directly from [Haasdonk and Ohlberger, 2011] or Theorem 4.4. □

This theorem establishes a bound for the norm of the model reduction error $x(t) - Vx_\mathrm{R}(t)$ as visualized in Figure 2.2. Therefore, model reduction methods that result in a good approximation of all states, such as POD presented in Section 2.1.2, are preferable compared to methods focusing on the I/O behavior, such as balanced truncation.

This error bound has several properties that ensure its computational efficiency. First, the error bound depends only on the initial condition of the detailed model as well as the state and input of the reduced model. Furthermore, the integral

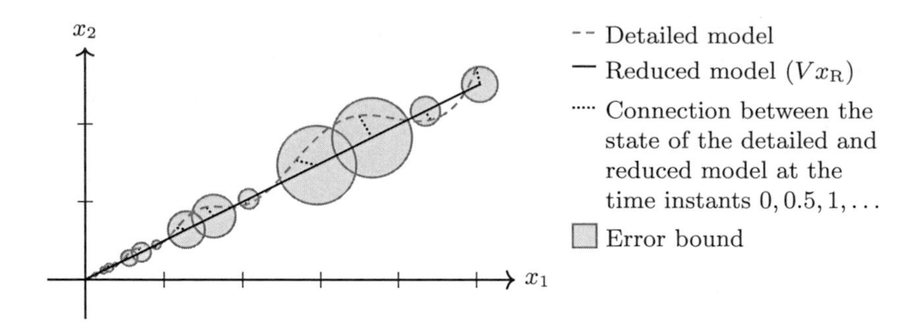

Figure 2.2: Visualization of the a-posteriori error bound.

in (2.5) does not have to be evaluated at every time t since the error bound $\Delta(t)$ can be computed by the scalar error bounding system

$$\dot{\Delta}(t) = \alpha\|r(t)\|\,, \qquad \Delta(0) = \alpha\|x(0) - Vx_\mathrm{R}(0)\|\,.$$

Finally, a direct computation of $\|r(t)\|$ can be computationally demanding, since the residual is a vector of dimension n. As discussed in [Haasdonk and Ohlberger, 2011],

$$\|r(t)\|^2 = x_\mathrm{R}^\mathsf{T}(t)M_1x_\mathrm{R}(t) + 2x_\mathrm{R}^\mathsf{T}(t)M_2u(t) + u^\mathsf{T}(t)M_3u(t)$$

with the matrices $M_1 = V^\mathsf{T}A^\mathsf{T}(I_n - VW^\mathsf{T})AV$, $M_2 = V^\mathsf{T}A^\mathsf{T}(I_n - VW^\mathsf{T})B$, and $M_3 = B^\mathsf{T}(I_n - VW^\mathsf{T})B$. Hence, the computation of $\|r(t)\|$ can be separated into an offline and online phase. In the offline phase, the matrices M_1, M_2, and M_3 are computed such that the computational complexity of the online phase is independent of the dimension of the detailed model. In [Haasdonk and Ohlberger, 2011], it is also shown that an increased computational efficiency can be achieved when using the reduced model with the error bound instead of the detailed model for simulation.

Using a scalar variable to bound the norm of the error as in (2.5) introduces conservatism. A reasonable tight error bound is attained in [Haasdonk and Ohlberger, 2011] by utilizing an appropriate order for the reduced model and a suitable norm $\|x\|_G := \sqrt{x^\mathsf{T}Gx}$ with a symmetric and positive definite matrix G. Furthermore, in [Ruiner et al., 2012], it is shown that a tighter bound is achieved by bounding the matrix exponential by a function of time, i.e., $\|\mathrm{e}^{At}\| \leq \alpha(t)$. Unfortunately, the resulting error bound is computationally expensive compared to the scalar error bounding system (2.5) since the convolution integral $\int_0^t \alpha(t-\tau)\|r(\tau)\|\,\mathrm{d}\tau$ has to be evaluated for all time instants t.

In Chapter 4, we provide an error bound, which improves the result in [Haasdonk and Ohlberger, 2011] in terms of conservatism but overcomes the computational demand of the approach in [Ruiner et al., 2012]. The proposed error bound is utilized in Chapters 5 and 6 to derive guarantees for MPC using reduced models. In the following section, we introduce the necessary background of MPC.

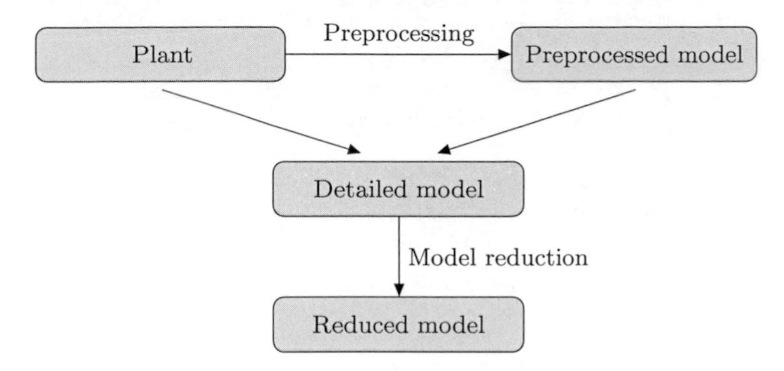

Figure 2.3: Denotation of the models used in this thesis.

2.3 Model Predictive Control

In this section, we introduce MPC and a common framework to guarantee convergence to the origin for MPC along the work in [Chen and Allgöwer, 1998; Findeisen et al., 2003; Fontes, 2001; Magni and Scattolini, 2004; Mayne et al., 2000]. Convergence to the origin is guaranteed for nominal MPC, i.e., the plant is equivalent to the prediction model. Hence, no uncertainty like disturbances or a model plant mismatch is present. Although MPC for linear systems is considered in Chapters 5 and 6, we present the framework for nonlinear systems as the cited work. Afterwards, we introduce an MPC approach using a reduced model. This approach neglects the model reduction error and, hence, the convergence is not guaranteed any more.

2.3.1 Problem Setup

We consider plants described by

$$\Sigma_{\mathrm{P}}: \begin{cases} \dot{x}_{\mathrm{P}}(t) = f_{\mathrm{P}}\big(x_{\mathrm{P}}(t), u_{\mathrm{P}}(t)\big)\,, \\ x_{\mathrm{P}}(0) = x_{\mathrm{P},0}\,, \end{cases} \tag{2.6}$$

in which $x_{\mathrm{P}}(t) \in \mathbb{R}^n$ is the state of the system at time $t \in \mathbb{R}_{0+}$, $u(t) \in \mathbb{R}^{n_{\mathrm{U}}}$ is the input vector at time $t \in \mathbb{R}_{0+}$, $u(\cdot) \in \mathcal{L}_2^{n_{\mathrm{U}}}$, and $x_{\mathrm{P},0} \in \mathbb{R}^n$ is the initial condition.

In this thesis, we distinguish between the plant model Σ_{P} and the detailed model Σ_{D}. In Chapters 4–6, the plant model is preprocessed. The resulting model is denoted preprocessed model. The model reduction is applied to the preprocessed model. In general, the detailed model, which is reduced, can be the plant model or a preprocessed model as visualized in Figure 2.3.

The states and inputs are constrained to a polytopic set by

$$C_{\mathrm{P}} \begin{bmatrix} x_{\mathrm{P}}(t) \\ u_{\mathrm{P}}(t) \end{bmatrix} \leq d_{\mathrm{P}}\,, \tag{2.7}$$

in which $C_{\mathrm{P}} \in \mathbb{R}^{n_{\mathrm{C}} \times n+n_{\mathrm{U}}}$ and $d_{\mathrm{P}} \in \mathbb{R}^{n_{\mathrm{C}}}$ for all $t \geq 0$. This includes, e.g., box constraints of the form

$$x_{\min} \leq x(t) \leq x_{\max}\,, \qquad u_{\min} \leq u(t) \leq u_{\max}\,,$$

in which $x_{\min} \in \mathbb{R}^n$, $x_{\max} \in \mathbb{R}^n$, $u_{\min} \in \mathbb{R}^{n_{\mathrm{U}}}$, and $u_{\max} \in \mathbb{R}^{n_{\mathrm{U}}}$.

Given the initial state $x_{\mathrm{P},0}$, we want to steer the system to the origin close to optimality with respect to the infinite horizon cost functional

$$J_{\mathrm{P}}^{\mathrm{inf}}(x_{\mathrm{P},0}, u_{\mathrm{P}}) := \int_0^\infty F_{\mathrm{P}}\big(x_{\mathrm{P}}(t), u_{\mathrm{P}}(t)\big)\, \mathrm{d}t\,,$$

subject to the system dynamics (2.6) as well as the state and input constraints (2.7). The quadratic stage cost is defined by

$$F_{\mathrm{P}}\big(x_{\mathrm{P}}(t), u_{\mathrm{P}}(t)\big) := x_{\mathrm{P}}^{\mathsf{T}}(t) Q_{\mathrm{P}} x_{\mathrm{P}}(t) + u_{\mathrm{P}}^{\mathsf{T}}(t) R_{\mathrm{P}} u_{\mathrm{P}}(t)\,. \tag{2.8}$$

The subsequent results rely on the following common assumptions (see, e.g., [Chen and Allgöwer, 1998; Findeisen et al., 2003]).

Assumption 2.2. *The vector field* $f_{\mathrm{P}} : \mathbb{R}^n \times \mathbb{R}^{n_{\mathrm{U}}} \to \mathbb{R}^n$ *is continuously differentiable and* $f_{\mathrm{P}}(0,0) = 0$.

Assumption 2.3. *The plant model* (2.6) *has a unique solution for all* $t \geq 0$, *for any initial condition* $x_{\mathrm{P},0} \in \mathbb{R}^n$, *and for any piecewise continuous and right-continuous input* $u_{\mathrm{P}}(\cdot) : \mathbb{R}_{0+} \to \mathbb{R}^{n_{\mathrm{U}}}$ *satisfying the constraints* (2.7).

Assumption 2.4. *The state and input constraints* (2.7) *define a compact set and the equilibrium* $x_{\mathrm{P}} = 0$ *with* $u_{\mathrm{P}} = 0$ *is contained in its interior.*

Assumption 2.5. *The matrices* Q_{P} *and* R_{P} *of the stage cost* (2.8) *are symmetric and positive definite.*

We use the definition of (asymptotic) stability according to [Khalil, 1996, Definition 3.1].

Definition 2.6. *The equilibrium* $(x_{\mathrm{P}}, u_{\mathrm{P}}) = (0,0)$ *of the system* (2.9) *is stable if for each* $\epsilon > 0$ *there is a* $\delta_\epsilon > 0$ *such that* $\|x_{\mathrm{P}}(0)\| < \delta_\epsilon$ *implies that* $\|x(t)\| < \epsilon$ *for all* $t \geq 0$. *The equilibrium is asymptotically stable if it is stable and* $\delta_\epsilon > 0$ *can be chosen such that* $\|x_{\mathrm{P}}(0)\| < \delta_\epsilon$ *implies that* $\lim_{t\to\infty} x_{\mathrm{P}}(t) = 0$.

When we say a system is (asymptotically) stable this means that the equilibrium at the origin for the unforced system is (asymptotically) stable.

One popular approach to tackle the given infinite horizon control problem is MPC.

2.3.2 Principle of Model Predictive Control

In MPC, at the sampling instant $t_i \geq 0$ the state $x_{\mathrm{P}}(t_i)$ is measured. Using the system dynamics, the model predictive controller predicts the system behavior over a finite prediction horizon T and computes an optimal input trajectory such that the state and inputs constraints are satisfied. To allow for a feedback, the optimal input is applied only until the next sampling instant t_{i+1}, where the whole procedure is repeated as depicted in Figure 2.4. For simplicity, we consider only equidistant sampling instants $t_i = i\delta$, $i \in \mathbb{N}_0$ with the sampling time $\delta > 0$.

The infinite horizon optimal control problem is approximated in MPC by a finite horizon optimal control problem. Due to the finite prediction horizon $T \geq \delta$, which is shifted from one sampling instant to the next, the predicted behavior can change as visualized in Figure 2.4. The resulting difference in the predicted and closed-loop behavior can lead to instability. By replacing the infinite prediction horizon with a finite one, MPC can destabilize, e.g., a stable four tank system as shown in [Raff et al., 2006].

The present MPC framework contains a terminal cost $E_{\mathcal{P}}(\cdot)$ and a terminal set $\Omega_{\mathcal{P}}$, which are used below in Section 2.3.3 to guarantee convergence of the closed-loop system despite the finite prediction horizon. The calligraphic subscript $\mathcal{P}$ denotes that the plant model is used for the prediction within the MPC scheme. Altogether, the following finite horizon optimal control problem is solved at each sampling instant $t_i = i\delta$, $i \in \mathbb{N}_0$.

Problem 2.7 (Optimization problem for MPC using the plant model).

$$
\begin{aligned}
\underset{\bar{u}_{\mathrm{P}} \in \mathcal{PC}^{n_{\mathrm{U}}}_{[t_i, t_i+T]}, \bar{x}_{\mathrm{P}}}{\text{minimize}} \quad & J_{\mathcal{P}}\big(x_{\mathrm{P}}(t_i), \bar{u}_{\mathrm{P}}(\cdot; t_i)\big) := \\
& \int_{t_i}^{t_i+T} F_{\mathrm{P}}\big(\bar{x}_{\mathrm{P}}(t; t_i), \bar{u}_{\mathrm{P}}(t; t_i)\big)\, \mathrm{d}t + E_{\mathcal{P}}\big(\bar{x}_{\mathrm{P}}(t_i + T; t_i)\big) \\
\text{subject to} \quad & \dot{\bar{x}}_{\mathrm{P}}(t; t_i) = f_{\mathrm{P}}\big(\bar{x}_{\mathrm{P}}(t; t_i), \bar{u}_{\mathrm{P}}(t; t_i)\big)\,, \\
& \bar{x}_{\mathrm{P}}(t_i; t_i) = x_{\mathrm{P}}(t_i)\,, \\
& C_{\mathrm{P}} \begin{bmatrix} \bar{x}_{\mathrm{P}}(t; t_i) \\ \bar{u}_{\mathrm{P}}(t; t_i) \end{bmatrix} \leq d_{\mathrm{P}}\,, \\
& \bar{x}_{\mathrm{P}}(t_i + T; t_i) \in \Omega_{\mathcal{P}}\,, \\
& \text{for all } t \in [t_i, t_i + T]\,.
\end{aligned}
$$

We denote by $\bar{x}_{\mathrm{P}}(\cdot; t_i)$ the predicted trajectory starting from initial condition $\bar{x}_{\mathrm{P}}(t_i; t_i) = x_{\mathrm{P}}(t_i)$ and driven by $\bar{u}_{\mathrm{P}}(t; t_i)$ for $t \in [t_i, t_i + T]$. We assume that the input trajectory that solves Problem 2.7 is given by $u^*_{\mathcal{P}}(t; t_i)$ with associated predicted state trajectory $x^*_{\mathcal{P}}(t; t_i)$ for $t \in [t_i, t_i + T]$. The MPC scheme using the plant model is visualized in Figure 2.5. The predicted variables $\bar{x}_{\mathrm{P}}(t; t_i)$, $\bar{u}_{\mathrm{P}}(t; t_i)$, $x^*_{\mathcal{P}}(t; t_i)$, and $u^*_{\mathcal{P}}(t; t_i)$ appear only inside of the model predictive controller and depend on the optimization time t_i. The optimal input trajectory $u^*_{\mathcal{P}}(\cdot; t_i)$ is

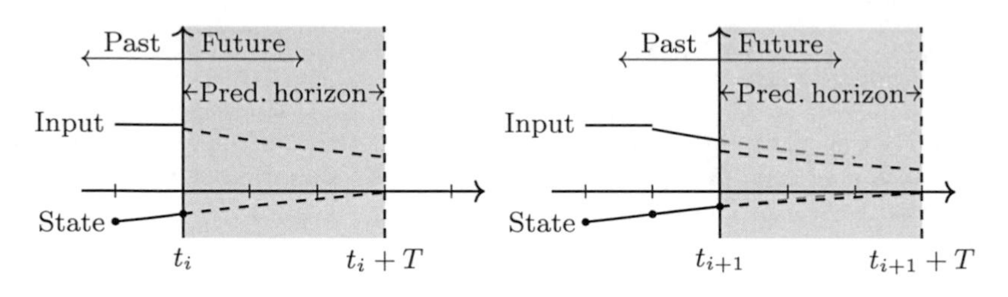

Figure 2.4: Idea of MPC. The gray dashed lines in the right figure are the trajectories predicted at the previous sampling instant.

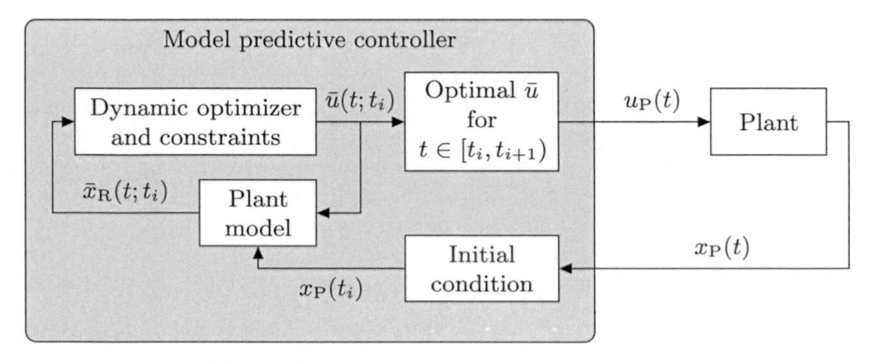

Figure 2.5: Structure of the $\mathcal{P}$-MPC scheme. The update of the sampling instant t_i is omitted.

applied to the plant until the next sampling instant $t_{i+1} = t_i + \delta$. Hence,

$$u_{\mathrm{P}}(t) = u_{\mathcal{P}}(t) := u^*_{\mathcal{P}}(t; t_i)\,, \qquad t_i \leq t < t_{i+1}\,. \tag{2.10}$$

The MPC scheme using the plant model is denoted $\mathcal{P}$-MPC and is given by the following algorithm.

Algorithm 2.8 ($\mathcal{P}$-MPC: MPC using the plant model).

Require: *The plant model* Σ_{P} (2.6), *the constraints* (2.7), *the sampling time* $0 < \delta \leq T$, *the matrices defining the stage cost* Q_{P}, R_{P}, *the terminal cost* $E_{\mathcal{P}}(\cdot)$, *and the terminal set* $\Omega_{\mathcal{P}}$

$i \leftarrow 0$

loop

$x_{\mathrm{P}}(t_i) \leftarrow$ *Measure the state of the plant at time* t_i

$u^*_{\mathcal{P}}(\cdot; t_i) \leftarrow$ *Solve Problem 2.7*

$u_{\mathrm{P}}(\cdot) = u_{\mathcal{P}}(\cdot) \leftarrow$ *Apply* $u^*_{\mathcal{P}}(t; t_i)$ *for* $t \in [t_i, t_{i+1})$ *to* Σ_{P}

$i \leftarrow i + 1$

end loop

Since $u_{\mathcal{P}}(\cdot)$ is defined by the repeated solution of Problem 2.7, feasibility of the optimization problem for all sampling instants has to be ensured in order to derive guarantees for the asymptotic behavior.

2.3.3 Guarantees in Model Predictive Control

To establish convergence to the origin of the plant controlled by MPC according to Algorithm 2.8 we state the following assumptions (see [Chen and Allgöwer, 1998; Findeisen et al., 2003]).

Assumption 2.9. *The terminal set $\Omega_{\mathcal{P}} \in \mathbb{R}^n$ is closed and contains the origin in its interior.*

Assumption 2.10. *The terminal cost is of the form $E_{\mathcal{P}}(x) := x^{\mathsf{T}} Q_{\mathcal{P}}^{\Omega} x$ and the matrix $Q_{\mathcal{P}}^{\Omega} \in \mathbb{R}^{n \times n}$ is symmetric and positive definite.*

Assumption 2.11. *There exists a local control law $u_{\mathrm{P}} = k(x_{\mathrm{P}})$ such that*

i) the terminal set $\Omega_{\mathcal{P}}$ is positive invariant for the plant model (2.6) *under the control law $u_{\mathrm{P}} = k(x_{\mathrm{P}})$,*

ii) for all $x_{\mathrm{P}} \in \Omega_{\mathcal{P}}$ the state and input constraints (2.7) *are satisfied for $u_{\mathrm{P}} = k(x_{\mathrm{P}})$, and*

iii) for all $x_{\mathrm{P}} \in \Omega_{\mathcal{P}}$ the closed loop with the plant model (2.6) *and the control law $u_{\mathrm{P}} = k(x_{\mathrm{P}})$ satisfies*

$$\dot{E}_{\mathcal{P}}(x_{\mathrm{P}}) + F_{\mathrm{P}}\big(x_{\mathrm{P}}, k(x_{\mathrm{P}})\big) \leq 0\,. \tag{2.11}$$

The positively invariant terminal set $\Omega_{\mathcal{P}}$ is often described by either a polytopic or an ellipsoidal sets. Polytopic positively invariant sets like the maximal constraint admissible set are arbitrarily complex and typically used for systems with a few states. Ellipsoidal sets are the only sets of practical use with a bounded complexity [Blanchini, 1999] and, hence, seem to be more appropriate than polytopic sets for complex control problems occurring frequently in industry.

Assumption 2.11 *i)* and *ii)* are used to ensure that from feasibility of Problem 2.7 at $t_0 = 0$ follows feasibility for all subsequent sampling instants when applying the model predictive controller. Furthermore, Assumption 2.11 guarantees that the terminal cost is an upper bound for the infinite horizon cost of the plant controlled by the local control law within the terminal set. This upper bound allows to guarantee convergence of the plant controlled by the model predictive controller (see [Chen and Allgöwer, 1998; Findeisen et al., 2003]).

Theorem 2.12 (Convergence of MPC using the plant model)**.** *Consider the plant Σ_{P}. Suppose that*

- *Assumptions 2.2–2.5, 2.9–2.11 are satisfied and*
- *Problem 2.7 is feasible at the sampling instant $t_0 = 0$.*

Then,

i) the plant (2.6) *in closed loop with the model predictive controller given by Algorithm 2.8 converges to the origin for $t \to \infty$.*

ii) The region of attraction of the closed loop is given by $\mathcal{F}_\mathcal{P}$, in which $\mathcal{F}_\mathcal{P} \subseteq \mathbb{R}^n$ denotes the set of all initial states $x(0)$ for which Problem 2.7 is feasible.

Proof. The theorem is proven, e.g., in [Chen and Allgöwer, 1998; Findeisen et al., 2003; Fontes, 2001; Magni and Scattolini, 2004]. □

2.3.4 Model Predictive Control Using Reduced Models

Typically, the major disadvantage of MPC is that the online optimization of a finite horizon optimal control problem is required. For high-dimensional systems, MPC typically leads to a prohibitive computational complexity and using reduced models for the prediction can serve as a remedy. As discussed in Section 1.2.3, the majority of literature about MPC using reduced models concentrates on the computational efficiency. The MPC schemes using reduced models proposed in Chapters 5 and 6 provide rigorous guarantees for the closed-loop behavior. These guarantees are achieved by predicting also a bound for the model reduction error, which unfortunately increases the computational complexity. Hence, it is interesting to compare the proposed MPC approaches with an approach neglecting the model reduction error. This MPC approach using a reduced model is introduced in this section.

The MPC approach using a reduced model emerges from the MPC formulation presented above by replacing the plant model with a reduced model. Furthermore, the predicted state of the plant is replaced with the estimated state $V\bar{x}_\mathrm{R}$. To initialize the reduced model the projected state of the plant is used. The MPC scheme using the reduced model is denoted $\mathcal{R}$-MPC and visualized in Figure 2.6.

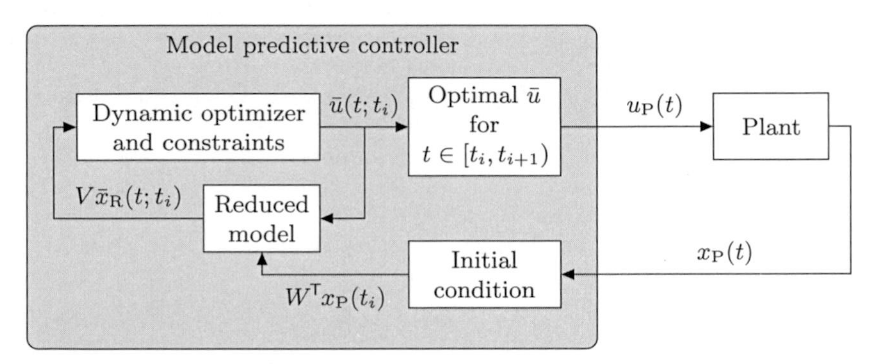

Figure 2.6: Structure of the $\mathcal{R}$-MPC scheme. The update of the sampling instant t_i is omitted.

At every sampling instant $t_i = i\delta$, $i \in \mathbb{N}_0$ the following finite horizon optimal control problem is solved.

Problem 2.13 (Optimization problem for MPC using the reduced model).

$$\underset{\bar{u}_\mathrm{P} \in \mathcal{PC}^{n_\mathrm{U}}_{[t_i, t_i+T]}, \bar{x}_\mathrm{R}}{\text{minimize}} \quad J_\mathcal{R}\big(x_\mathrm{P}(t_i), \bar{u}_\mathrm{P}(\cdot; t_i)\big) := \int_{t_i}^{t_i+T} F_\mathrm{P}\big(V\bar{x}_\mathrm{R}(t;t_i), \bar{u}_\mathrm{P}(t;t_i)\big)\,\mathrm{d}t + E_\mathcal{P}\big(V\bar{x}_\mathrm{R}(t_i+T;t_i)\big)$$

subject to

$$\dot{\bar{x}}_\mathrm{R}(t;t_i) = A_\mathrm{R}\bar{x}_\mathrm{R}(t;t_i) + B_\mathrm{R}\bar{u}_\mathrm{P}(t;t_i)\,, \tag{2.12a}$$

$$\bar{x}_\mathrm{R}(t_i;t_i) = W^\mathsf{T} x_\mathrm{P}(t_i)\,, \tag{2.12b}$$

$$C_\mathrm{P}\begin{bmatrix} V\bar{x}_\mathrm{R}(t;t_i) \\ \bar{u}_\mathrm{P}(t;t_i) \end{bmatrix} \leq d_\mathrm{P}\,, \tag{2.12c}$$

$$V\bar{x}_\mathrm{R}(t_i+T;t_i) \in \Omega_\mathcal{P}\,, \tag{2.12d}$$

$$\text{for all } t \in [t_i, t_i+T]\,.$$

We assume that the input trajectory that solves Problem 2.13 is given by $u^*_\mathcal{R}(t;t_i)$ with associated predicted state trajectory $x^*_\mathcal{R}(t;t_i)$ for $t \in [t_i, t_i+T]$. The calligraphic subscript $\mathcal{R}$ denotes that the reduced model is used for the prediction. The optimal input trajectory $u^*_\mathcal{R}(\cdot;t_i)$ is applied to the plant until the next sampling instant $t_{i+1} = t_i + \delta$. Hence,

$$u_\mathrm{P}(t) = u_\mathcal{R}(t) := u^*_\mathcal{R}(t;t_i)\,, \qquad t_i \leq t < t_{i+1}\,. \tag{2.13}$$

The MPC scheme using the reduced model is denoted $\mathcal{R}$-MPC and is given by the following algorithm.

Algorithm 2.14 ($\mathcal{R}$-MPC: MPC using the reduced model).

Require: *The plant model* Σ_P (2.6)*, the matrices* V *and* W *defining the reduced model* Σ_R (2.12a)–(2.12b)*, the constraints* (2.7)*, the sampling time* $0 < \delta \leq T$*, the matrices defining the stage cost* Q_P*,* R_P*, the terminal cost* $E_\mathcal{P}(\cdot)$*, and the terminal set* $\Omega_\mathcal{P}$

$i \leftarrow 0$

loop

$x_\mathrm{P}(t_i) \leftarrow$ *Measure the state of the plant at time* t_i

$u^*_\mathcal{R}(\cdot;t_i) \leftarrow$ *Solve Problem 2.13*

$u_\mathrm{P}(\cdot) = u_\mathcal{R}(\cdot) \leftarrow$ *Apply* $u^*_\mathcal{R}(t;t_i)$ *for* $t \in [t_i, t_{i+1})$ *to* Σ_P

$i \leftarrow i+1$

end loop

$\mathcal{R}$-MPC is a common approach in MPC using reduced models, found similarly for discrete-time systems, e.g., in [Agudelo et al., 2007a; Hovland et al., 2008a; Marquez et al., 2013; Narciso and Pistikopoulos, 2008]. The cited references focus on different aspects and therefore, among others, differences in the terminal set and the initial condition of the reduced model exist.

By using a reduced model for the prediction, the model reduction error is introduced within the MPC scheme. Thus, guarantees for the closed loop with $\mathcal{P}$-MPC are not necessarily valid any more. For instance, we cannot conclude from the state and input constraints (2.12c) that the constraints (2.7) are satisfied. Furthermore, since the predicted state of the reduced model and the projected measured state of the plant are in general different, Problem 2.13 can get infeasible after some time or the closed loop can be unstable when $\mathcal{R}$-MPC is used to control the plant although $\mathcal{P}$-MPC stabilizes the plant. This is shown in Chapter 5 by applying $\mathcal{P}$-MPC and $\mathcal{R}$-MPC to a motivating example. Furthermore, $\mathcal{R}$-MPC is used as a reference for the proposed MPC approaches with respect to the computation time. For this purpose, an example motivated from real world is used, which is introduced in the following section.

2.4 Nonisothermal Tubular Chemical Reactor

To evaluate the error bound and the MPC schemes proposed in Chapters 4–6, the model of a nonisothermal tubular chemical reactor is utilized. Tubular reactors are commonly used in the chemical industry since they are relatively easy to maintain, produce usually the highest conversion per reactor volume of any of the flow reactors, and allow for large-scale and low cost production [Agudelo, 2009; Fogler, 2009]. Unfortunately, it is difficult to control the temperature inside the reactor and hot spots can occur in the case of exothermic reactions [Fogler, 2009]. Besides the practical relevance, another reason to consider tubular reactors is the work in [Agudelo et al., 2007a,b], where two MPC schemes using a reduced model are discussed in order to control a tubular reactor. The following problem setup is based largely on [Agudelo et al., 2007a]. Slight modifications of the setup considered in [Agudelo et al., 2007a] are necessary since both controllers in [Agudelo et al., 2007a,b] do neither guarantee satisfaction of state constraints nor asymptotic stability. For example, the stage cost has to be modified since the setpoint in [Agudelo et al., 2007a] is not a steady state and, hence, Assumptions 2.2–2.5 are not satisfied.

In this thesis, we consider a tubular reactor consisting of a cylindrical pipe, which is fed by a continuous flow with a concentration $C_{\text{in}}(t) \in \mathbb{R}$ of the reactant R and fluid temperature $T_{\text{in}}(t) \in \mathbb{R}$. While the fluid flows from the inlet to the outlet, a first order, irreversible, exothermic reaction takes place in which the desired product P is produced. In order to prevent undesired byproducts, the fluid temperature inside the reactor must be kept smaller than 400 K. To control the fluid temperature, the temperatures $T_{\text{J}}(t) = \begin{bmatrix} T_{\text{J},1} & T_{\text{J},2} & T_{\text{J},3} \end{bmatrix}^\mathsf{T} \in \mathbb{R}^3$ of three cooling/heating jackets can be changed independently between 280 K and 400 K as visualized in Figure 2.7.

Due to variations in an upstream plant the fluid temperature at the inlet $T_{\text{in}}(\cdot)$ may change from the nominal value $T_{\text{nom}} = 340\,\text{K}$ to a value within $[315\,\text{K}, 365\,\text{K}]$. The inlet reactant concentration $C_{\text{in}}(t)$ is equal to the nominal value $C_{\text{nom}} =$

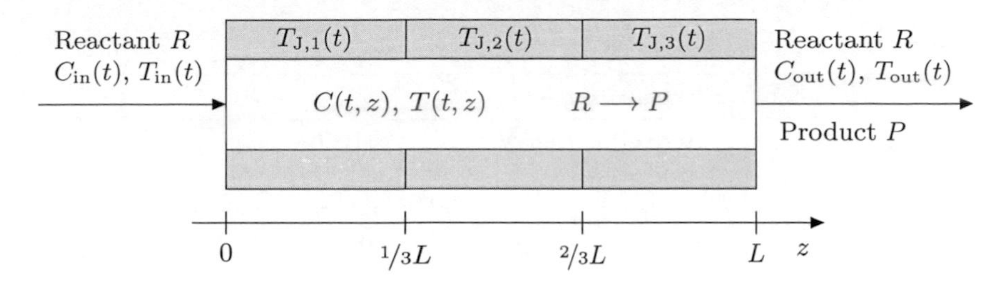

Figure 2.7: Tubular reactor with three cooling/heating jackets. The schematic is based on [Agudelo et al., 2007a].

$0.02\,\text{mol/l}$ for all t. In the following, we consider only step changes of $T_{\text{in}}(t)$ at $t = 0$, i.e.,

$$T_{\text{in}}(t) := \begin{cases} T_{\text{nom}} & \text{if } t < 0\,, \\ T_{\text{in,const}} & \text{if } t \geq 0\,, \end{cases}$$

with $T_{\text{in,const}} \in [315\,\text{K}, 365\,\text{K}]$. Since tubular reactors are almost always operated at steady state [Fogler, 2009], we assume that the reactor is at the steady state for the nominal inlet temperature at $t = 0$. The control goal is to achieve a high conversion ratio despite these step changes of $T_{\text{in}}(\cdot)$ while satisfying the temperature constraints.

2.4.1 Derivation of the Plant Model

The considered model of the tubular reactor assumes plug-flow behavior, i.e., without radial variations in the velocity, concentration, and temperature. Hence, only variations along the axial coordinate, on the spatial domain $z \in [0\,\text{m}, L]$, exist. Furthermore, diffusion and dispersion as well as heat transfer effects between the reactor wall and the jacket fluids are neglected. Therefore, the reactor wall temperature $T_{\text{w}}(t,z)$ is equal to the temperature of the jackets, i.e.,

$$T_{\text{w}}(t,z) = \begin{cases} T_{\text{J},1}(t) & \text{if } 0\,\text{m} \leq z < 1/3L\,, \\ T_{\text{J},2}(t) & \text{if } 1/3L \leq z < 2/3L\,, \\ T_{\text{J},3}(t) & \text{if } 2/3L \leq z \leq L\,. \end{cases}$$

We denote the concentration of the reactant with $C(t,z)$ and the temperature inside the reactor with $T(t,z)$. With the given assumptions, the tubular reactor is

Table 2.1: Parameters of the tubular reactor [Agudelo et al., 2007b].

Parameter	Explanation	Value
E	Activation energy	$11250\,{}^{\text{cal}}/_{\text{mol}}$
$G_\text{r} = \frac{-\Delta H\,k_0}{\rho\,C_\text{P}}$		$4.25 \cdot 10^9\,{}^{\text{l}\cdot\text{K}}/_{\text{mol}\cdot\text{s}}$
$H_\text{r} = \frac{4\,h}{d\,\rho\,C_\text{P}}$		$0.2\,{}^1/_\text{s}$
k_0	Kinetic constant	$10^6\,{}^1/_\text{s}$
L	Reactor length	$1\,\text{m}$
R_gas	Ideal gas constant	$1.986\,{}^{\text{cal}}/_{\text{mol}\cdot\text{K}}$
v	Fluid superficial velocity	$0.1\,{}^\text{m}/_\text{s}$

described by two coupled nonlinear PDEs [Agudelo et al., 2007a]

$$\frac{\partial C}{\partial t}(t,z) = -v\frac{\partial C(t,z)}{\partial z} - k_0\,C(t,z)\exp\left(-\frac{E}{R_\text{gas}T(t,z)}\right), \tag{2.14a}$$

$$\frac{\partial T}{\partial t}(t,z) = -v\frac{\partial T(t,z)}{\partial z} + G_\text{r}\,C(t,z)\exp\left(-\frac{E}{R_\text{gas}T(t,z)}\right) + H_\text{r}\big(T_\text{w}(t,z) - T(t,z)\big) \tag{2.14b}$$

with the boundary conditions

$$C(t,0) = C_\text{in}(t) \quad \text{and} \quad T(t,0) = T_\text{in}(t)$$

and initial conditions

$$C(0,z) = C_0(z) \quad \text{and} \quad T(0,z) = T_0(z)\,.$$

The values of the parameters are given in Table 2.1. As in [Agudelo et al., 2007a,b], these values are taken from [Smets et al., 2002], which are themselves inspired by [Fjeld and Ursin, 1971].

The PDE model (2.14) is linearized around the nominal inlet fluid temperature, the nominal inlet reactant concentration, the jacket temperatures $T_{\text{J,lin}}$, and the resulting steady state profiles for the reactant concentration $C_\text{lin}(\cdot)$ and fluid temperature $T_\text{lin}(\cdot)$ determined in [Agudelo et al., 2007b]. After the linearization, the reactor is divided into 150 sections with a step size $\Delta z = {}^1/_{150}\,\text{m}$ and the partial derivatives are approximated by backward differences. Details of the linearization and spatial discretization can be found in Appendix A and [Agudelo et al., 2007b]. In this section, we merely present the resulting LTI ODE model. For this purpose, we define the vectors for the concentrations and temperatures at the grid points $i\,\Delta z$, $i = 1, \ldots, 150$,

$$C_\text{grid}(t) := \begin{bmatrix} C(t,\Delta z) \\ \vdots \\ C(t, 150\,\Delta z) \end{bmatrix} \in \mathbb{R}^{150}, \qquad T_\text{grid}(t) := \begin{bmatrix} T(t,\Delta z) \\ \vdots \\ T(t, 150\,\Delta z) \end{bmatrix} \in \mathbb{R}^{150},$$

as well as the state vector and the linearization point

$$x(t) := \begin{bmatrix} \frac{C_{\mathrm{grid}}(t)}{C_{\mathrm{nom}}} \\ \frac{T_{\mathrm{grid}}(t)}{T_{\mathrm{nom}}} \\ \frac{T_{\mathrm{J}}(t)}{T_{\mathrm{nom}}} \end{bmatrix} \in \mathbb{R}^{303}, \qquad x_{\mathrm{lin}} := \begin{bmatrix} C_{\mathrm{lin}}(\Delta z)/C_{\mathrm{nom}} \\ \vdots \\ C_{\mathrm{lin}}(150\,\Delta z)/C_{\mathrm{nom}} \\ T_{\mathrm{lin}}(\Delta z)/T_{\mathrm{nom}} \\ \vdots \\ T_{\mathrm{lin}}(150\,\Delta z)/T_{\mathrm{nom}} \\ T_{\mathrm{J,lin}}/T_{\mathrm{nom}} \end{bmatrix} \in \mathbb{R}^{303},$$

and the input vector consisting of the normalized time derivatives of the jacket temperatures

$$u(t) := \dot{T}_{\mathrm{J}}(t)/T_{\mathrm{nom}} \in \mathbb{R}^{3}.$$

The time derivatives of the jacket temperatures are considered as inputs instead of the jacket temperatures to allow for a penalization of the time derivatives of the jacket temperatures in the control objective. The ODE model resulting from the linearization and spatial discretization can be written in the form

$$\dot{x}(t) = A\big(x(t) - x_{\mathrm{lin}}\big) + Bu(t) + B_{\mathrm{T}}\big(T_{\mathrm{in}}(t) - T_{\mathrm{nom}}\big)/T_{\mathrm{nom}}, \qquad x(0) = x_0. \tag{2.15}$$

The matrices A and B as well as the vector B_{T} are stated in Appendix A. With respect to the state vector $x(t) - x_{\mathrm{lin}}$ and the external inputs $u(t)$ and $\big(T_{\mathrm{in}}(t) - T_{\mathrm{nom}}\big)/T_{\mathrm{nom}}$ the model is LTI.

The constraints for the fluid and jacket temperatures mentioned on page 30 have to be satisfied at all the grid points $i\,\Delta z$, $i = 1, \dots, 150$. Thus, the constraints for the LTI model are

$$T_{\mathrm{grid}}(t) \leq 400\,\mathrm{K}, \tag{2.16a}$$

$$280\,\mathrm{K} \leq T_{\mathrm{J}}(t) \leq 400\,\mathrm{K}. \tag{2.16b}$$

The modeling error arising from the linearization and the spatial discretization are not considered in this thesis, since we focus on the error originating from the model reduction. In Chapters 4–6, we only consider the ODE model (2.15), not the original PDE, and study the proposed error bound and MPC schemes by this practically motivated example.

2.4.2 Problem Setup for Model Predictive Control

To apply the MPC schemes discussed in this thesis to the tubular reactor, the control goal has to be specified appropriately. As mentioned on page 31, the control goal for the tubular reactor is to achieve a high product concentration at the outlet

$$C_{\mathrm{out}}(t) = C(t, L) = C(t, N\,\Delta z)$$

despite the step change of $T_{\text{in}}(t)$ at $t = 0$ while satisfying the temperature constraints. To achieve this objective, we use an approach that is often used in process control [Rawlings and Mayne, 2009]. First, an optimal setpoint is determined and, second, MPC is used to steer the system to this setpoint.

The inflow temperature dependent setpoint $\big(x_{\text{SP}}(T_{\text{in}}), u_{\text{SP}}(T_{\text{in}})\big)$ with

$$x_{\text{SP}}(T_{\text{in}}) = \begin{bmatrix} C_{\text{SP}}(T_{\text{in}})/C_{\text{nom}} \\ T_{\text{SP}}(T_{\text{in}})/T_{\text{nom}} \\ T_{\text{J,SP}}(T_{\text{in}})/T_{\text{nom}} \end{bmatrix} \in \mathbb{R}^{150+150+3}, \qquad u_{\text{SP}}(T_{\text{in}}) = 0 \in \mathbb{R}^3$$

is determined by the solution of the following optimization problem.

Problem 2.15 (Optimization problem to determine the setpoint)**.**

$$\begin{aligned} \underset{x_{\text{SP}}(T_{\text{in}})}{\text{minimize}} \quad & 0.7\left(\frac{C_{\text{SP},150}(T_{\text{in}})}{C_{\text{nom}}}\right)^2 + \frac{0.3}{150}\sum_{i=1}^{150}\left(\frac{T_{\text{SP},i}(T_{\text{in}}) - T_{\text{nom}}}{T_{\text{nom}}}\right)^2 \\ \text{subject to} \quad & 0 = A\big(x_{\text{SP}}(T_{\text{in}}) - x_{\text{lin}}\big) + B_T\left(\frac{T_{\text{in}} - T_{\text{nom}}}{T_{\text{nom}}}\right), && \text{(2.17a)} \\ & T_{\text{SP}}(T_{\text{in}}) \leq 395\,\text{K}, && \text{(2.17b)} \\ & 285\,\text{K} \leq T_{\text{J,SP}}(T_{\text{in}}) \leq 395\,\text{K}. && \text{(2.17c)} \end{aligned}$$

The first summand in the cost function penalizes the concentration at the outlet while the second summand weights the deviation from the nominal temperature. In contrast to [Agudelo et al., 2007a,b], the model predictive controllers in this thesis guarantee satisfaction of the constraints. Thus, we allow a setpoint closer to the temperature constraint (2.16a). Furthermore, we choose a higher weight for the output concentration as considered in [Logist et al., 2005]. Hence, higher temperature deviations from T_{nom} occur, consequently, making the temperature constraint (2.16a) more important. Altogether, the output concentration at the setpoint for the nominal inflow is $5.1 \cdot 10^{-4}$ mol/l and, compared to [Agudelo et al., 2007a,b], it is decreased by a factor of three.

The resulting setpoints of the tubular reactor for different inlet fluid temperatures are visualized in Figure 2.8. For small T_{in}, the largest possible jacket temperature $T_{\text{J},1}(T_{\text{in}}) = 395\,\text{K}$ is used to heat up the fluid. The temperature of the second jacket is chosen smaller in order to satisfy the temperature constraint (2.17b). To achieve a low concentration of the reactant at the outlet for small T_{in}, a high $T_{\text{J},3}(T_{\text{in}})$ is chosen. Hence, the temperature constraint (2.17b) is active also for some $z \in [0.67\,\text{m}, 1\,\text{m}]$ if $T_{\text{in}} < 320.1\,\text{K}$ and, therefore, $T_{\text{J},3}(T_{\text{in}})$ is limited by the temperature constraint (2.17b). For large T_{in} the reactant concentration is smaller. Hence, there is less heat from the reaction for $z \geq 0.33\,\text{m}$ and the temperature constraint (2.17b) is inactive for $T_{\text{in}} \geq 357.23\,\text{K}$. Summarizing, the setpoint depends nonlinearly on T_{in} due to the underlying optimization problem.

With (2.17a), the ODE model (2.15) can be written as

$$\dot{x}(t) = A\big(x(t) - x_{\text{SP}}(T_{\text{in}})\big) + Bu(t), \qquad x(0) = x_0 = x_{\text{SP}}(T_{\text{nom}}).$$

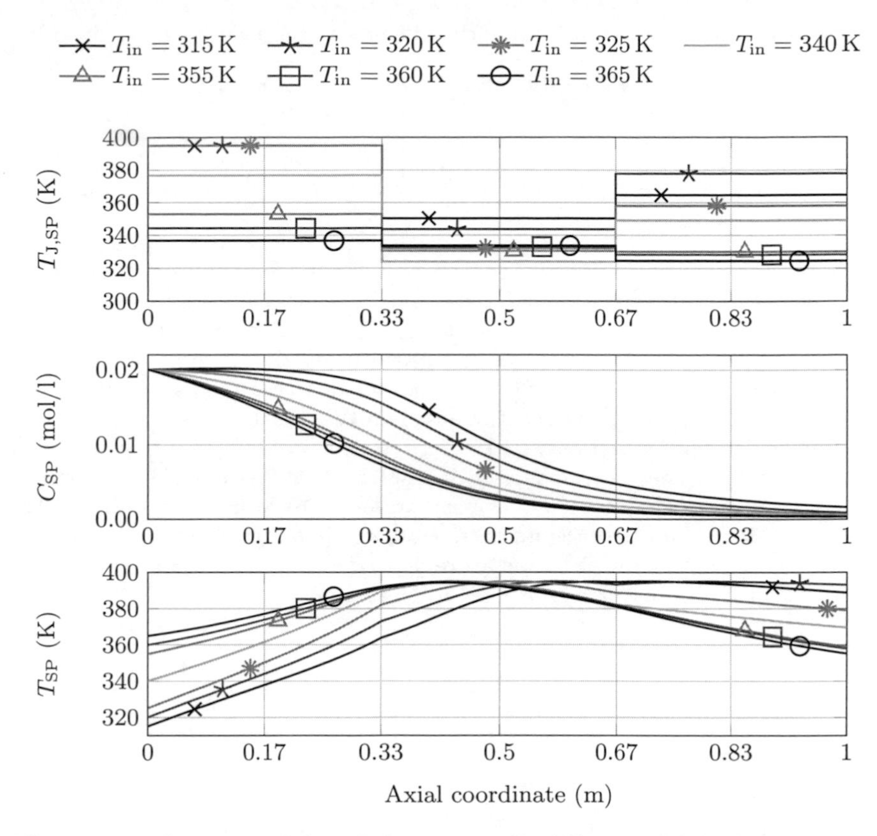

Figure 2.8: Setpoints of the tubular reactor for different inlet fluid temperatures and the steady state profiles used for the linearization.

Hence, the externally given T_{in} is entirely encapsulated in the setpoint. A shift of the origin to $\big(x_{\text{SP}}(T_{\text{in}}), u_{\text{SP}}(T_{\text{in}})\big) = \big(x_{\text{SP}}(T_{\text{in}}), 0\big)$ and using that $T_{\text{in}}(t)$ is constant for $t \geq 0$ results in

$$\frac{\mathrm{d}\Delta x(t)}{\mathrm{d}t} = A\,\Delta x(t) + Bu(t)\,, \qquad \Delta x(0) = x_{\text{SP}}(T_{\text{nom}}) - x_{\text{SP}}(T_{\text{in}})\,, \tag{2.18}$$

in which $\Delta x(t) = x(t) - x_{\text{SP}}(T_{\text{in}})$. Hence, a nominal MPC scheme like the one presented in Section 2.3 can be applied to steer the system to the setpoint. For the LTI model (2.18), Assumptions 2.2 and 2.3 are satisfied. By assuming further state and input constraints in addition to (2.16), Assumption 2.4 can also be satisfied.

The quadratic stage cost weighting the difference to the setpoint is defined by

$$F(x,u) := 10^{-5}\left\|\frac{C_{\text{grid}}(t) - C_{\text{SP}}(T_{\text{in}})}{C_{\text{nom}}}\right\|^2 + 10\left(\frac{C_{\text{out}}(t) - C_{\text{SP},150}(T_{\text{in}})}{C_{\text{nom}}}\right)^2$$
$$+ 117\left\|\frac{T_{\text{grid}}(t) - T_{\text{SP}}(T_{\text{in}})}{T_{\text{nom}}}\right\|^2 + 0.25\left\|\frac{T_{\text{J}}(t) - T_{\text{J,SP}}(T_{\text{in}})}{T_{\text{nom}}}\right\|^2 + 1.1\left\|\frac{\dot{T}_{\text{J}}(t)}{T_{\text{nom}}}\right\|^2. \tag{2.19}$$

Thus, deviations in the output concentration are strongly penalized. The values weighting $C_{\text{out}}(t) = C_{\text{grid},150}$, $T_{\text{grid}}(t)$, and $\dot{T}_{\text{J}}(t)$ are derived from [Agudelo et al., 2007a]. The other two terms weighting $C_{\text{grid}}(t)$ and $T_{\text{J}}(t)$ are added to satisfy Assumption 2.5 and to force the jacket temperatures to converge to the setpoint.

The plant and cost function for the discrete-time case used in Chapter 6 is computed with the matrix exponential based on [Van Loan, 1978].

Summarizing, in this section we have derived an LTI ODE model of a tubular reactor along the lines of [Agudelo et al., 2007b]. For this model, a practically motivated control problem setup has been presented, in which the controller has to deal with step changes of the uncontrollable inlet fluid temperature. By using a shift of the origin to an inlet temperature dependent setpoint, the given problem setup is transformed into a nominal MPC problem.

Chapter 3

Trajectory-Based Model Reduction for Nonlinear Systems

Methods for model reduction of nonlinear systems are much less developed than for linear system as discussed in Section 1.2.4. In this chapter, we present a model reduction method for nonlinear continuous-time dynamical systems, which allows to obtain models of low order and low computational complexity. In contrast to many existing model reduction methods, the method relies on I/O trajectories of the detailed model and not on a projection. Hence, we named the method trajectory-based model reduction. For simplicity, the I/O is skipped.

After the problem statement in Section 3.1, the basic procedure of the trajectory-based model reduction is described in Section 3.2. This basic procedure is applied to the model of an MAPK cascade in Section 3.2.5. In Section 3.3, the procedure is extended to preserve the location and local exponential stability of multiple steady states. Furthermore, the developed procedure is applied to the Fermi-Pasta-Ulam lattice. The presented approach is compared on a conceptual level with existing approaches relying only on simulated trajectories in Section 3.4.

The content of this chapter is very similar to [Löhning et al., 2011a,b]. The main difference is the comparison with other approaches in Section 3.4.

3.1 Problem Statement

In this chapter, we consider the problem of model reduction for systems of nonlinear ODEs. The system to be reduced is called detailed model and abbreviated using the subscript D. For notational simplicity the detailed model is assumed to be a single-input single-output system,

$$\Sigma_{\mathrm{D}} : \begin{cases} \dot{x}(t; x_0, u) = f(x(t; x_0, u), u(t)) \,, & x(0) = x_0 \,, \\ y_{\mathrm{D}}(t; x_0, u) = h(x(t; x_0, u), u(t)) \,, \end{cases} \tag{3.1}$$

in which $x(t; x_0, u) \in \mathbb{R}^n$ is the state of the system, $u(t) \in \mathbb{R}$ is the input, and $y_{\mathrm{D}}(t; x_0, u) \in \mathbb{R}$ is the output at time $t \in \mathbb{R}_{0+}$ and n is the order of the model. To ensure existence and uniqueness of solutions the vector field $f : \mathbb{R}^n \times \mathbb{R} \to \mathbb{R}^n$ is assumed to be globally Lipschitz continuous. In addition to Section 2.1, only n_{R} times continuous differentiable inputs $u \in \mathcal{L}^1_\infty[0, T_{\mathrm{end}}]$ are allowed, in which n_{R} is the order of the reduced model. Furthermore, the mappings f and $h : \mathbb{R}^n \times \mathbb{R} \to \mathbb{R}$ are considered to be sufficiently smooth.

Given a detailed model Σ_D, we are interested in a reduced model Σ_R that provides a good approximation of the I/O behavior of Σ_D on a finite time interval $[0, T_\mathrm{end}]$. To allow for the development of problem-specific and thus smaller reduced models, we allow for the integration of prior knowledge about relevant initial conditions and input trajectories. Such prior knowledge will in the following be encoded in a weighting function $p_{\mathrm{x},\varphi}$ assigning importance to the initial condition and the input trajectory. For ease of presentation, the input trajectories u_φ are assumed to be parameterized by $\varphi \in \mathbb{R}^{n_\varphi}$, e.g., a finite Fourier series, so it is possible to express the weighting as a function of the parameters φ.

For the weighting function $p_{\mathrm{x},\varphi} : \mathbb{R}^n \times \mathbb{R}^{n_\varphi} \to \mathbb{R}_{0+}$ we assume that the integral $\int\int p_{\mathrm{x},\varphi}(x_0, \varphi)\,\mathrm{d}x_0\,\mathrm{d}\varphi$ exists and is equal to 1. Based on this, $p_{\mathrm{x},\varphi}(x_0, \varphi)$ can be interpreted as a probability density, in which $p_{\mathrm{x},\varphi}(x_0, \varphi)$ represents the probability of obtaining x_0 as initial condition and u_φ as input trajectory.

The weighting function is a design variable that allows for a tailored reduced model. For example, if the reduced model is used for control, the input trajectories should represent the trajectories occurring in closed loop [Marquardt, 2002, page 36]. In [Löhning et al., 2011a], it is suggested to use two independent weighting functions, one for the initial condition and one for the input trajectory. In joint work with Slusarek [2012] it emerged that the two independent weighting functions can prevent a precise definition of the I/O behavior that should be approximated by the reduced model. For example, consider a detailed model with limit cycles that depend on the constant input value. Furthermore, we are interested in a reduced model that approximates these input dependent limit cycles. If these limit cycles do not have at least one common point in the state space, the I/O behavior that can be defined by two independent weighting functions for the initial condition and the input trajectory contains in addition to the input dependent limit cycles also the transients from the initial condition to the input dependent limit cycles. These transients can be avoided with one common weighting function.

The following remark summarizes parts of the joint work with Slusarek [2012] and gives a guideline on how to choose the importance weight.

Remark 3.1 (Defining the important I/O behavior)**.** *If the order of the reduced model is chosen smaller than the order of the detailed model, it is implicitly assumed that the essential behavior of the detailed model can be approximately described by an n_R-dimensional manifold — compare to the nonlinear Galerkin method [Matthies and Meyer, 2003]. Hence, it is important that the relevant trajectories of the detailed model evolve close to this low-dimensional manifold. Consequently, the weighting of the initial conditions have to be chosen carefully. For instance, the initial conditions should be restricted such that the deviation from a low-dimensional manifold is small in comparison to the excitation of the state by the input trajectories.*

While performing model reduction, one main goal in addition to a small approximation error is to preserve the key features of the detailed model. Two of these key properties are the location and the stability of steady states. Therefore, we assume that the set of locally exponentially stable steady states of the detailed model denoted by $\mathcal{S} := \{(u_{\mathrm{SS}}^{(s)}, y_{\mathrm{SS}}^{(s)})\}_{s=1}^{n_{\mathrm{S}}}$, is available. Since I/O based model reduction is considered, we only specify the steady states via a constant output $y_{\mathrm{SS}}^{(s)}$. Thus, we allow for internal dynamics, e.g., due to limit cycle oscillations that are not observable at the output.

Based on this setup, we consider the following problem.

Problem 3.2. *Given a detailed model Σ_{D} and the relevant I/O behavior defined by the weighting function $p_{\mathrm{x},\varphi}(x_0, \varphi)$, compute a reduced model*

$$\Sigma_{\mathrm{R}} : \begin{cases} \dot{x}_{\mathrm{R}}(t; x_0, u_\varphi) = f_{\mathrm{R}}(x_{\mathrm{R}}(t; x_0, u_\varphi), u_\varphi(t)), & x_{\mathrm{R}}(0) = g_{\mathrm{R}}(x_0), \\ y_{\mathrm{R}}(t; x_0, u_\varphi) = h_{\mathrm{R}}(x_{\mathrm{R}}(t; x_0, u_\varphi), u_\varphi(t)), & \end{cases}$$

of fixed order n_{R} such that
1) the objective functional

$$J(\Sigma_{\mathrm{R}}) := \iint E(x_0, u_\varphi)\, p_{\mathrm{x},\varphi}(x_0, \varphi)\, \mathrm{d}x_0\, \mathrm{d}\varphi$$

is minimized, in which E is the integrated squared output error

$$E(x_0, u_\varphi) := \int_0^{T_{\mathrm{end}}} \left(y_{\mathrm{D}}(t; x_0, u_\varphi) - y_{\mathrm{R}}(t; x_0, u_\varphi)\right)^2 \mathrm{d}t$$

and
2) the reduced model Σ_{R} has the locally exponentially stable steady states given by the set $\mathcal{S}$.

In short, the goal is to find, for a given order n_{R} of the reduced model, the vector field f_{R}, the output mapping h_{R} and the initial state mapping g_{R}, such that the weighted integrated squared output error is minimal and the location and stability of the steady states in $\mathcal{S}$ is preserved.

3.2 Procedure of Trajectory-Based Model Reduction

In this section, we present a novel procedure to address the first part of Problem 3.2. Afterwards, the second part of Problem 3.2 is incorporated in Section 3.3.

This section starts with the evaluation of the objective functional in Section 3.2.1. Afterwards, the observability normal form is used to parameterize the reduced model in Section 3.2.2. A convex optimization problem that determines the parameters of the reduced model, is deduced in Section 3.2.3. In Section 3.2.4, a sparsity enhancing ℓ_1-norm formulation is used to reduce the computational complexity of the reduced model. The proposed procedure is utilized to reduce the nonlinear model of the MAPK cascade in Section 3.2.5.

3.2.1 Evaluation of the Objective Functional

The basic idea of trajectory-based model reduction is the usage of simulated trajectories for the assessment of the reduced model. Accordingly, the objective functional $J(\Sigma_\mathrm{R})$ is determined using simulation. Since the integral defining $J(\Sigma_\mathrm{R})$ is high-dimensional $(1+n+n_\varphi)$, classical integration methods are not applicable.

In this work, we use Monte-Carlo integration [MacKay, 2005] to overcome this problem. To determine an approximation of the objective functional $J(\Sigma_\mathrm{R})$, a sample of initial conditions $\{x_0^{(i)}\}_{i=1}^{n_\mathrm{T}}$ and input trajectories $\{u_\varphi^{(i)}\}_{i=1}^{n_\mathrm{T}}$ is drawn, in which $u_\varphi^{(i)} := u_{\varphi^{(i)}}$. Furthermore, $\left(x_0^{(i)}, \varphi^{(i)}\right) \sim p_{\mathrm{x},\varphi}(x_0, \varphi)$. The notation $z^{(i)} \sim p_z(z)$ denotes that the sample member $z^{(i)}$ is drawn according to $p_z(z)$. The number of sample members is denoted by n_T. Furthermore, a sample of time points $\{t^{(i)}\}_{i=1}^{n_\mathrm{T}}$ is drawn from a uniform distribution over $[0, T_\mathrm{end}]$. Given $\{x_0^{(i)}\}_{i=1}^{n_\mathrm{T}}$, $\{u_\varphi^{(i)}\}_{i=1}^{n_\mathrm{T}}$, and $\{t^{(i)}\}_{i=1}^{n_\mathrm{T}}$, a sample of points from I/O trajectories is given by

$$\mathcal{T} := \left\{ y_\mathrm{D}^{(i)}\left(t^{(i)}; x_0^{(i)}, u_\varphi^{(i)}\right), u_\varphi^{(i)}\left(t^{(i)}\right) \right\}_{i=1}^{n_\mathrm{T}} .$$

From classical Monte-Carlo integration it is now known that for $n_\mathrm{T} \gg 1$ a good approximation of $J(\Sigma_\mathrm{R})$ is

$$\hat{J}(\Sigma_\mathrm{R}) := \frac{1}{n_\mathrm{T}} \sum_{i=1}^{n_\mathrm{T}} \left(y_\mathrm{D}^{(i)}\left(t^{(i)}; x_0^{(i)}, u_\varphi^{(i)}\right) - y_\mathrm{R}^{(i)}\left(t^{(i)}; x_0^{(i)}, u_\varphi^{(i)}\right) \right)^2 .$$

Given the approximated, evaluable objective functional $\hat{J}(\Sigma_\mathrm{R})$, we are left with the problem of minimizing $\hat{J}(\Sigma_\mathrm{R})$ over f_R, h_R, and g_R.

3.2.2 Parameterization of the Reduced Model

In order to keep the optimization of the reduced model tractable, we assume that the reduced model Σ_R is globally observable. This assumption is crucial but not restrictive, since it ensures that the reduced model has no hidden states.

Given that the reduced model Σ_R is globally observable and only n_R times continuous differentiable inputs u_φ are allowed, each possible reduced model can be represented using its observability normal form [Zeitz, 1984]

$$\dot{\xi}_j = \xi_{j+1}, \qquad j = 1, \ldots, n_\mathrm{R} - 1, \tag{3.2a}$$

$$\dot{\xi}_{n_\mathrm{R}} = \nu(\xi, \mu), \tag{3.2b}$$

$$y_\mathrm{R} = \xi_1, \tag{3.2c}$$

with the extended input vector $\mu := \begin{bmatrix} u_\varphi & \dot{u}_\varphi & \ldots & \frac{\mathrm{d}^{n_\mathrm{R}}}{\mathrm{d}t^{n_\mathrm{R}}} u_\varphi \end{bmatrix}^\mathsf{T} \in \mathbb{R}^{n_\mathrm{R}+1}$ and function $\nu : \mathbb{R}^{n_\mathrm{R}} \times \mathbb{R}^{n_\mathrm{R}+1} \to \mathbb{R}$. In this unique state representation, the states of the reduced model are the time derivatives of the output, i.e.,

$$\xi = \begin{bmatrix} y_\mathrm{R} & \dot{y}_\mathrm{R} & \ldots & \frac{\mathrm{d}^{n_\mathrm{R}-1}}{\mathrm{d}t^{n_\mathrm{R}-1}} y_\mathrm{R} \end{bmatrix}^\mathsf{T} \in \mathbb{R}^{n_\mathrm{R}} .$$

This is beneficial since also a direct link between the states ξ and the output of the detailed model is established, which makes a reasonable choice for the mapping of the initial states g_{R},

$$\xi_k(0) := \frac{\mathrm{d}^{k-1}}{\mathrm{d}t^{k-1}} y_{\mathrm{D}}\big(0; x_0, u_\varphi\big)\,, \qquad k = 1, \ldots, n_{\mathrm{R}}\,.$$

Representation (3.2) facilitates the identification of a reduced model. Instead of identifying the vector-valued functions f_{R} and g_{R} and the scalar-valued function h_{R}, we are left with the problem of identifying a single scalar-valued function ν. Nevertheless, identifying ν still requires an optimization over a function space.

To obtain a parameterized model, the right-hand side of $\dot{\xi}_{n_{\mathrm{R}}}$ is restricted to a weighted sum of n_θ ansatz functions

$$\nu_\theta(\xi, \mu) := \theta^{\mathsf{T}} m(\xi, \mu)\,.$$

We denote the weighting parameters by $\theta \in \mathbb{R}^{n_\theta}$ and the ansatz functions by $m(\xi, \mu) =: \begin{bmatrix} m_1(\xi, \mu) & \ldots & m_{n_\theta}(\xi, \mu) \end{bmatrix}^{\mathsf{T}}$. Common ansatz functions are polynomials or radial basis functions.

Given the parameterized function ν_θ, the problem of estimating Σ_{R} is reduced to estimating θ. Hence, the cost functional $\hat{J}(\Sigma_{\mathrm{R}})$ is only a function of θ.

3.2.3 Estimation of the Reduced Model

Nonlinear Optimization Approach

One method to estimate the parameters θ is nonlinear optimization. This approach is the predominant approach for nonlinear continuous-time parameter estimation.

In this model reduction context, this classical approach requires to solve the following optimization problem.

Problem 3.3 (Nonlinear optimization approach).

$$\underset{\theta, \xi}{\text{minimize}} \quad \frac{1}{n_{\mathrm{T}}} \sum_{i=1}^{n_{\mathrm{T}}} \Big(y_{\mathrm{D}}^{(i)}\big(t^{(i)}; x_0^{(i)}, u_\varphi^{(i)}\big) - y_{\mathrm{R}}^{(i)}\big(t^{(i)}; x_0^{(i)}, u_\varphi^{(i)}\big) \Big)^2 ,$$

$$\text{subject to} \quad \dot{\xi}_j^{(i)} = \xi_{j+1}^{(i)}\,, \qquad j = 1, \ldots, n_{\mathrm{R}} - 1\,, \tag{3.3a}$$

$$\dot{\xi}_{n_{\mathrm{R}}}^{(i)} = \theta^{\mathsf{T}} m\big(\xi^{(i)}, \mu^{(i)}\big)\,, \tag{3.3b}$$

$$\xi_k^{(i)}(0) = \frac{\mathrm{d}^{k-1}}{\mathrm{d}t^{k-1}} y_{\mathrm{D}}^{(i)}\big(0; x_0^{(i)}, u_\varphi^{(i)}\big)\,, \qquad k = 1, \ldots, n_{\mathrm{R}}\,, \tag{3.3c}$$

$$y_{\mathrm{R}}^{(i)} = \xi_1^{(i)}\,, \tag{3.3d}$$

for all $i = 1, \ldots, n_{\mathrm{T}}$.

Since this problem is in general nonlinear and nonconvex, computation of the global optimum cannot be ensured. Additionally, for algorithms seeking the global optimum the computational effort grows in general tremendously with the number of sample members n_{T}.

Convex Optimization Approach

In this thesis, another approach is proposed that exploits the fact that model reduction is performed and not classical parameter estimation from measurement data.

While performing model reduction, not only the output y_D but also its derivatives are available. Considering the reduced model, this means that the target values for all states ξ of the reduced model are known. In particular, we have a target value for $\dot{\xi}_{n_\mathrm{R}}$, the n_R-th derivative of the output y_R. If we could ensure that

$$\forall t, \forall x_0, \forall u_\varphi : \frac{\mathrm{d}^{n_\mathrm{R}}}{\mathrm{d}t^{n_\mathrm{R}}} y_\mathrm{R}(t; x_0, u_\varphi) = \frac{\mathrm{d}^{n_\mathrm{R}}}{\mathrm{d}t^{n_\mathrm{R}}} y_\mathrm{D}(t; x_0, u_\varphi),$$

is satisfied, the I/O mapping of the detailed and reduced model would match exactly. This inspires a modification of the objective functional to

$$\hat{J}_m(\theta) := \frac{1}{n_\mathrm{T}} \sum_{i=1}^{n_\mathrm{T}} \left(\frac{\mathrm{d}^{n_\mathrm{R}}}{\mathrm{d}t^{n_\mathrm{R}}} y_\mathrm{D}^{(i)}\left(t^{(i)}; x_0^{(i)}, u_\varphi^{(i)}\right) - \theta^\mathsf{T} m\left(\xi_\mathrm{D}^{(i)}, \mu^{(i)}\right)\right)^2, \tag{3.4}$$

with

$$\xi_\mathrm{D}^{(i)} := \left[y_\mathrm{D}^{(i)}\left(t^{(i)}; x_0^{(i)}, u_\varphi^{(i)}\right), \dots, \frac{\mathrm{d}^{n_\mathrm{R}-1}}{\mathrm{d}t^{n_\mathrm{R}-1}} y_\mathrm{D}^{(i)}\left(t^{(i)}; x_0^{(i)}, u_\varphi^{(i)}\right)\right]^\mathsf{T},$$

$$\mu^{(i)} := \left[u_\varphi^{(i)}\left(t^{(i)}\right), \dots, \frac{\mathrm{d}^{n_\mathrm{R}}}{\mathrm{d}t^{n_\mathrm{R}}} u_\varphi^{(i)}\left(t^{(i)}\right)\right]^\mathsf{T}.$$

This modified objective function $\hat{J}_m(\theta)$ is closely related to $\hat{J}(\theta)$. The main difference is that $\hat{J}_m(\theta)$ does not penalize the integrated error along the trajectories. Instead, only the difference in the change of the output of the detailed and reduced model at the time points $t^{(i)}$ is penalized. This results in a minimization of the short-term prediction error. The modified error criterion $\frac{\mathrm{d}^{n_\mathrm{R}}}{\mathrm{d}t^{n_\mathrm{R}}} y_\mathrm{D}^{(i)}(t^{(i)}; x_0^{(i)}, u_\varphi^{(i)}) - \theta^\mathsf{T} m(\xi_\mathrm{D}^{(i)}, \mu^{(i)})$ is comparable to the 1-step-ahead prediction error of an autoregressive model with exogenous input in discrete-time systems identification. Hence, in analogy to discrete-time systems identification [Ljung, 1999], we say that the modified objective function $\hat{J}_m(\theta)$ minimizes the equation error.

With the modification of the objective functional we obtain the following optimization problem.

Problem 3.4 (Convex optimization approach).

$$\underset{\theta}{\text{minimize}} \quad \hat{J}_m(\theta) = \frac{1}{n_\mathrm{T}} \sum_{i=1}^{n_\mathrm{T}} \left(\frac{\mathrm{d}^{n_\mathrm{R}}}{\mathrm{d}t^{n_\mathrm{R}}} y_\mathrm{D}^{(i)}\left(t^{(i)}; x_0^{(i)}, u_\varphi^{(i)}\right) - \theta^\mathsf{T} m\left(\xi_\mathrm{D}^{(i)}, \mu^{(i)}\right)\right)^2.$$

Solving Problem 3.4 is equivalent to computing the parameterization θ of ν_θ that results in the smallest difference $\frac{\mathrm{d}^{n_\mathrm{R}}}{\mathrm{d}t^{n_\mathrm{R}}} y_\mathrm{D} - \frac{\mathrm{d}^{n_\mathrm{R}}}{\mathrm{d}t^{n_\mathrm{R}}} y_\mathrm{R}$ assuming that the reduced model is in the correct point in the space of outputs and derivatives thereof.

Compared to the nonlinear optimization approach in Problem 3.3, the Problem 3.4 has one big advantage. By defining

$$M := \left[m(\xi_{\mathrm{D}}^{(1)}, \mu^{(1)}) \quad \dots \quad m(\xi_{\mathrm{D}}^{(n_{\mathrm{T}})}, \mu^{(n_{\mathrm{T}})})\right],$$

$$\psi := \left[\frac{\mathrm{d}^{n_{\mathrm{R}}}}{\mathrm{d}t^{n_{\mathrm{R}}}} y_{\mathrm{D}}^{(1)}\left(t^{(1)}; x_0^{(1)}, u_\varphi^{(1)}\right) \quad \dots \quad \frac{\mathrm{d}^{n_{\mathrm{R}}}}{\mathrm{d}t^{n_{\mathrm{R}}}} y_{\mathrm{D}}^{(n_{\mathrm{T}})}\left(t^{(n_{\mathrm{T}})}; x_0^{(n_{\mathrm{T}})}, u_\varphi^{(n_{\mathrm{T}})}\right)\right],$$

we can rewrite Problem 3.4 to

$$\underset{\theta}{\text{minimize}} \quad \frac{1}{n_{\mathrm{T}}} \left\| \theta^{\mathrm{T}} M - \psi \right\|^2 .$$

Since M and ψ are determined in advance, Problem 3.4 is a convex quadratic optimization problem.

Remark 3.5. *Computing the solution of Problem 3.4 is independent of the detailed model, assuming the sample of points $\mathcal{T}$ has been computed beforehand. This emphasizes that only the complexity of the I/O behavior of the detailed model is relevant for the proposed trajectory-based model reduction procedure. Thus, this model reduction procedure can be applied to large-scale detailed models. Unfortunately, the order of the reduced model is limited due to numerical difficulties originating from the model structure with the derivatives of the output as states, i.e., the observability normal form.*

3.2.4 Complexity Reduction of the Reduced Model

Although Problem 3.4 allows for an efficient computation of θ, one problem remains. In general, all elements of θ are nonzero, which results in a reduced model with n_θ ansatz functions. Such a complex reduced model is often unnecessary. It is desirable to include only the required monomials by forcing θ to have many zero elements.

To enforce sparsity of θ a method known from compressive sensing [Candès et al., 2006; Donoho, 2006] can be used. In this work, an ℓ_1-norm formulation is used, which results in sparse solutions. With W being the diagonal matrix with the ℓ_2-norm of the rows of M on the diagonal, $\theta = W^{-1}\tilde{\theta}$, and θ^* the solution of Problem 3.4, the corresponding convex optimization problem is

Problem 3.6 (Sparsity enhancing optimization approach).

$$\begin{aligned} &\underset{\tilde{\theta}}{\text{minimize}} && \|\tilde{\theta}\|_1 \\ &\text{subject to} && \hat{J}_m\left(W^{-1}\tilde{\theta}\right) \leq \gamma \hat{J}_m(\theta^*) . \end{aligned}$$

The normalization with W ensures that the elements of $\tilde{\theta}$ are comparable. The parameter $\gamma \geq 1$ restricts the increase of the objective function and thus the decrease in approximation accuracy. Hence, the choice of γ is critical and the trade off between model complexity and approximation quality has to be kept in mind.

To summarize, we presented an approach to derive reduced models in observability normal form by convexifying Problem 3.2. Using ℓ_1-minimization the complexity of the resulting models can be further reduced.

3.2.5 Example: MAPK Cascade

In this section, some important properties of the proposed model reduction approach are illustrated by studying the mitogen-activated protein kinase (MAPK) cascade. The MAPK cascade is one of the most intensively studied pathways in eukaryotic cells [Orton et al., 2005; Schoeberl et al., 2002]. It plays an important role in many signaling processes and is crucially involved in different cell fate decisions such as growth, proliferation, differentiation, and cell survival.

The MAPK cascade has been chosen due to its importance in the field of system biology and mainly the I/O behavior is of interest, making it perfectly suited for our approach. Furthermore, a major property of the MAPK cascade is the existence of limit cycles. This nonlinear behavior indicates that a nonlinear model reduction procedure is required.

MAPK Cascade Model

For this study, the model proposed by Kholodenko [2000] has been chosen because it is well established. The model has eight state variables, which are the concentration of the three kinases Raf, MEK, and ERK in different phosphorylation states. The input of the MAPK cascade is the activity of the enzyme Ras,

$$u_\varphi(t) = c \cdot [\text{Ras}](t)\,,$$

which catalyzes the phosphorylation of Raf. The factor c is a normalization constant ensuring $u_\varphi(t) \in [0, 1]$. The output is the concentration of double phosphorylated ERK

$$y_\text{D}(t) = [\text{ERK-PP}](t)\,.$$

The differential equations governing the time evolution of the state variable can be found in Appendix B. All reaction rates are modeled by first or second-order Hill kinetics, which are rational functions. The unit for the time is min and nmol for the output y. The units are omitted in the following.

Model Reduction

To perform the model reduction, the data is generated by simulating the detailed MAPK cascade model. The weighting function for the initial condition and the input trajectory is chosen to be the product of two independent weighting functions, i.e., $p_{\text{x},\varphi}(x_0, \varphi) = p_\text{x}(x_0)p_\varphi(\varphi)$. This choice is in contrast to Remark 3.1. We choose the weighting functions as in [Löhning et al., 2011a] in this thesis since the benefits of choosing the initial condition in dependence of the input trajectory are already presented in Slusarek [2012] for the MAPK cascade. The weighting function for the initial condition $p_x(x_0)$ is the sum of two Gaussian distributions. The mean values of these Gaussian distributions are the centers of the limit cycles that are exhibited by the system for $u_\varphi = 0.1$ and $u_\varphi = 1$. The standard deviations of the initial conditions are 5% and 10% of the upper limit of the states. Due to conservation relations the initial conditions are restricted to a linear subspace with equal absolute amount of the three kinases, e.g., Raf + Raf-P = Raf_{tot}. The input trajectories are chosen to be step functions

$$\breve{u}_\varphi(t) := \begin{cases} \tilde{u}_1 & \text{if } t \leq \tau\,, \\ \tilde{u}_2 & \text{if } t > \tau \end{cases}$$

smoothed by the transfer function $G(s) = 1/(100s+1)^{n_{\text{R}}}$,

$$u_\varphi(t) = \mathcal{L}^{-1}\big(G(s)\big) * \breve{u}_\varphi(t) + \tilde{u}_1 \mathcal{L}^{-1}\left(\frac{1-G(s)}{s}\right), \tag{3.5}$$

with the convolution operator $*$, Laplace transform $\mathcal{L}$, and the parameters $\varphi = \begin{bmatrix}\tilde{u}_1 & \tilde{u}_2 & \tau\end{bmatrix}^\mathsf{T}$. The second summand in (3.5) ensures that $u_\varphi(0) = \tilde{u}_1$. The weighting function for the input is

$$p_\varphi(\varphi) := \Phi(\tilde{u}_1 \mid 0,1)\, \Phi(\tilde{u}_2 \mid 0,1)\, \Phi(\tau \mid 0,150)\,,$$

in which $\Phi(x \mid a,b)$ denotes the probability density of the uniform distribution in the interval $x \in [a,b]$. For the model reduction a sample with 2000 members is drawn from $p_x(x_0)$, $p_\varphi(\varphi)$, and $p_t(t) := \Phi(t \mid 0,150)$ respectively. For the resulting pairs of x_0 and φ the MAPK cascade is simulated, yielding the sample of I/O trajectories $\mathcal{T}$.

Given the sample $\mathcal{T}$, the order n_{R} of the reduced model has to be chosen. Since the detailed model has a relative degree three, n_{R} is set to three. Due to the nonlinear kinetics of the detailed model, the ansatz functions are chosen to be all monomials with degree less or equal five, i.e.,

$$m(\xi,\mu) := \begin{bmatrix}1 & \xi_1 & \ldots & \mu_{n_{\text{R}}+1} & \xi_1^2 & \xi_1\xi_2 & \ldots & \mu_{n_{\text{R}}+1}^5\end{bmatrix}^\mathsf{T}.$$

This yields the reduced model with up to $n_\theta = \binom{2n_{\text{R}}+6}{5} = 792$ parameters and monomials,

$$\Sigma_{\text{R}} : \begin{cases} \dot{\xi}_1 = \xi_2\,, \\ \dot{\xi}_2 = \xi_3\,, \\ \dot{\xi}_3 = \nu_\theta(\xi, u_\varphi) = \theta^\mathsf{T} m(\xi, u_\varphi)\,, \\ \xi_k(0) = \dfrac{\mathrm{d}^{k-1}}{\mathrm{d}t^{k-1}} y_{\text{D}}\big(0; x_0, u_\varphi\big)\,, \qquad k \in \{1,2,3\}\,, \\ y_{\text{R}} = \xi_1\,. \end{cases}$$

Based on this structure of the reduced model and the sample $\mathcal{T}$, a parameterization of the reduced model is determined by solving Problem 3.6 using the MATLAB

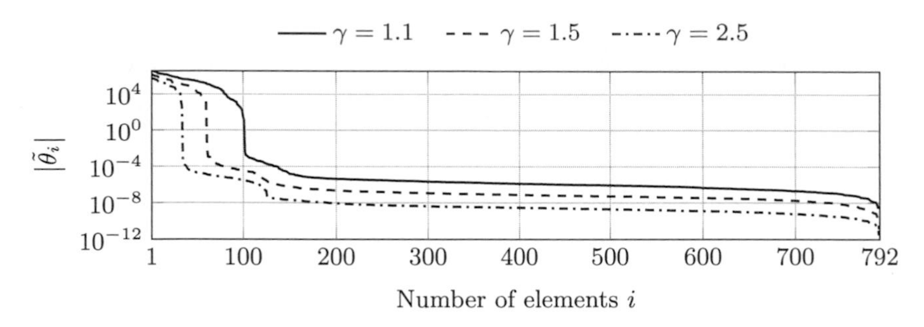

Figure 3.1: Dependency of the absolute values of the coefficients $|\tilde{\theta}_i|$ on the allowed increase of the objective function. The elements $|\tilde{\theta}_i|$ are sorted in decreasing order.

toolbox `Yalmip` [Löfberg, 2009] with the solver `SeDuMi` [Sturm, 1999]. By changing the optimization parameter γ, we can thereby reduce the number of numerical nonzero elements in θ, as depicted in Figure 3.1. A trade-off between model accuracy and complexity is achieved with $\gamma = 1.5$, which results in 61 monomials. The large number of required monomials is partially due to the rational functions contained in the original model.

Validation of the Reduced Model

To validate the reduced model, the approximation error of the third derivative of the output is evaluated. Figure 3.2 depicts the standard deviation of the absolute value of the approximation error $\left|\dddot{y}_{\mathrm{D}}^{(i)} - \theta^{\mathsf{T}} m\big(\xi_{\mathrm{D}}^{(i)}, \mu^{(i)}\big)\right|$ and $\left|\theta^{\mathsf{T}} m\big(\xi_{\mathrm{D}}^{(i)}, \mu^{(i)}\big)\right|$ for a validation sample consisting of 2000 members with $x_0^{(i)} \sim p_{\mathrm{x}}(x_0)$ and $\varphi^{(i)} = \begin{bmatrix}\tilde{u}_1^{(i)} & \tilde{u}_2^{(i)} & \tau^{(i)}\end{bmatrix}^{\mathsf{T}}$ with $\tilde{u}_1^{(i)} \sim \Phi(\tilde{u}_1 \mid 0, 1)$, $\tilde{u}_2^{(i)} \sim \Phi(\tilde{u}_2 \mid 0, 1)$, and $\tau^{(i)} = 75$. The standard deviation of the approximation error decays rapidly. Furthermore, it is more than one order of magnitude smaller than the standard deviation of $\theta^{\mathsf{T}} m\big(\xi_{\mathrm{D}}^{(i)}, \mu^{(i)}\big)$, when the model is not excited. The larger values after $t = 0$, respectively $t = 75$ are due to the initial conditions and/or input signals. The convergence of the trajectories to the nonlinear manifold defined by $\dddot{y}_{\mathrm{D}} - \theta^{\mathsf{T}} m\big(\xi_{\mathrm{D}}^{(i)}, \mu^{(i)}\big) = 0$ suggests that the long term behavior can be reproduced well.

The difference of the output trajectory between the detailed and reduced model is not directly related to $\dddot{y}_{\mathrm{D}} - \theta^{\mathsf{T}} m\big(\xi_{\mathrm{D}}^{(i)}, \mu^{(i)}\big)$. Thus, trajectories from simulations with $x_0 \sim p_{\mathrm{x}}(x_0)$ are used as further validation. Exemplary, trajectories for an increasing and a decreasing step as input signal are shown in Figure 3.3. The quantitative behavior is reproduced well by the reduced model. To emphasize the necessity of a nonlinear model reduction procedure, we also compute a linear model with three states using the proposed procedure, i.e., $m(\xi, \mu) = \begin{bmatrix}\xi_1 & \xi_2 & \xi_3 & u_\varphi & \dot{u}_\varphi & \ddot{u}_\varphi & \dddot{u}_\varphi\end{bmatrix}$. This yields a large decrease in performance as can be seen in Figure 3.3.

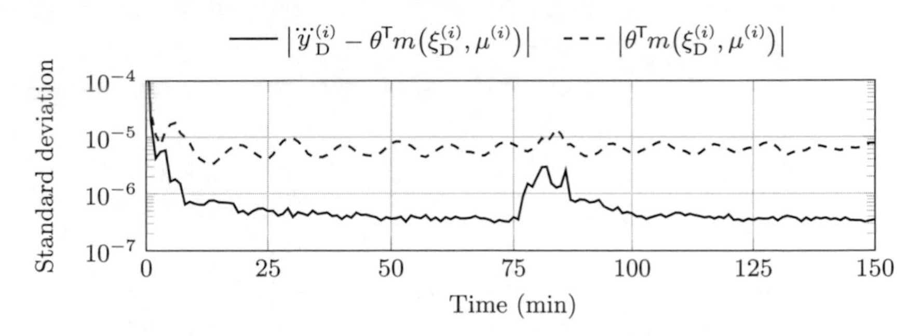

Figure 3.2: Standard deviation of the absolute error in the third derivative of the output $\left|\dddot{y}_{\mathrm{D}}^{(i)} - \theta^{\mathsf{T}} m(\xi_{\mathrm{D}}^{(i)}, \mu^{(i)})\right|$ and the standard deviation of the absolute value of the third derivative of the output captured by the reduced model $\left|\theta^{\mathsf{T}} m(\xi_{\mathrm{D}}^{(i)}, \mu^{(i)})\right|$ as a function of time for a validation data set.

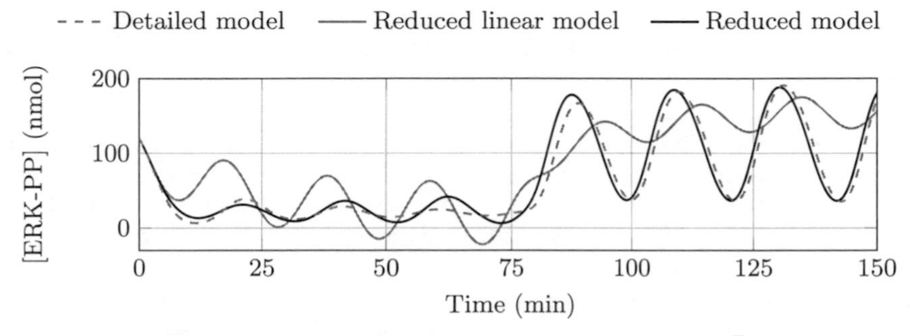

(a) Output trajectories for an increasing step with $\varphi = \begin{bmatrix} 0.1 & 1 & 75 \end{bmatrix}^{\mathsf{T}}$ as input signal.

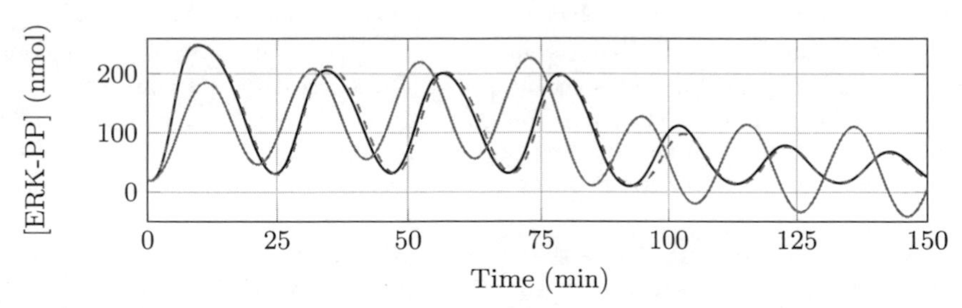

(b) Output trajectories for a decreasing step with $\varphi = \begin{bmatrix} 0.9 & 0.2 & 75 \end{bmatrix}^{\mathsf{T}}$ as input signal.

Figure 3.3: Trajectories of the detailed, reduced linear, and reduced model.

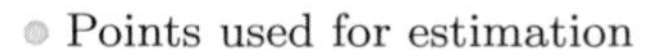

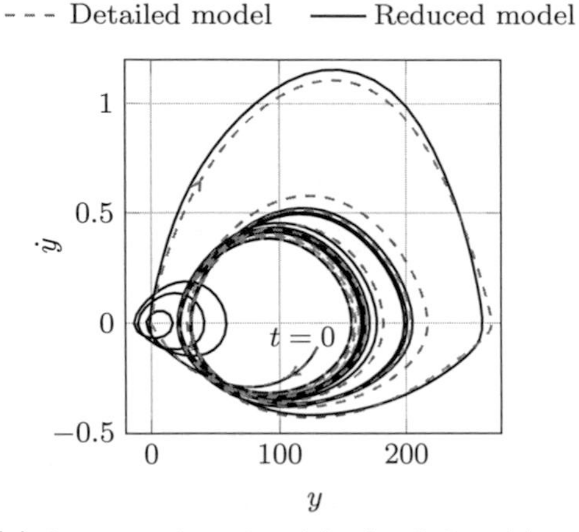

(a) Output trajectories of the detailed model and the reduced model for $t \in [0, 1500]$, and $\varphi = \begin{bmatrix} 0 & 0.9 & 75 \end{bmatrix}^\mathsf{T}$ projected on the y-$\dot{y}$-plane.

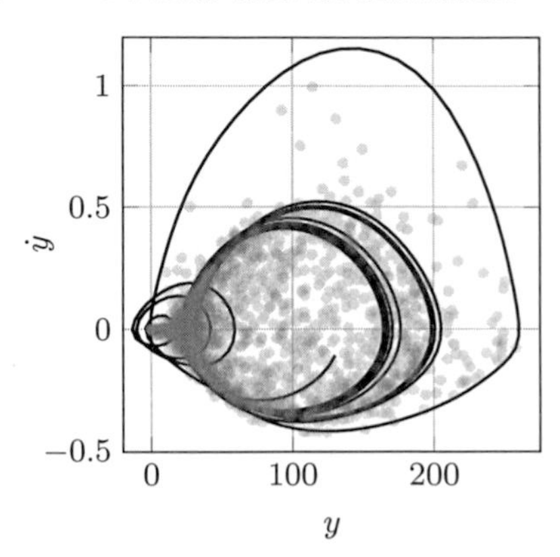

(b) Points used for estimation of the mapping $\theta^\mathsf{T} m(\xi, \mu)$.

Figure 3.4: Trajectories in the y-$\dot{y}$-plane.

The existence of limit cycles is a major property of the MAPK cascade model, stressing its nonlinearity. To visualize the limit cycle, the trajectories of the outputs for $t \in [0, 1500]$ and $\varphi = \begin{bmatrix} 0 & 0.9 & 75 \end{bmatrix}^\mathsf{T}$, projected onto the y-$\dot{y}$-plane, are shown in Figure 3.4a. The limit cycle is approximately captured by the reduced model.

Besides the limit cycle, Figure 3.4b shows the points used for estimation of the mapping $\theta^\mathsf{T} m(\xi, \mu)$ projected onto the y-$\dot{y}$-plane. An analysis of this set unravels that the trajectory of y_D for $u_\varphi = 0$ is close to the border of supporting points. This makes the estimation of the mapping $\theta^\mathsf{T} m(\xi, \mu)$ difficult around $(y_\mathrm{D}, \dot{y}_\mathrm{D}) = (0, 0)$. One result of this is that the reduced model returns negative values for y_R.

To determine how well the reduced model reproduces the limit cycles for different values of $\tilde{u}_2$, Figure 3.5 shows the amplitude and frequency of the limit cycle. Thereby, the amplitude is defined by $0.5\big(\max(y) - \min(y)\big)$, in which y is the trajectory of the limit cycle, and the frequency is defined by the inverse of the cycle duration. Figure 3.5 shows a good agreement between the detailed and reduced model.

The preceding analysis shows that the model reduction procedure leads to a reduced model that captures the quantitative behavior of the MAPK cascade model for a wide range of input trajectories. The detailed model of the MAPK cascade with eight states and rational functions could be reduced to a polynomial model with three states and 61 parameters. This indicates that on the one hand redundancies in the model, arising from the conservation relations of the kinases,

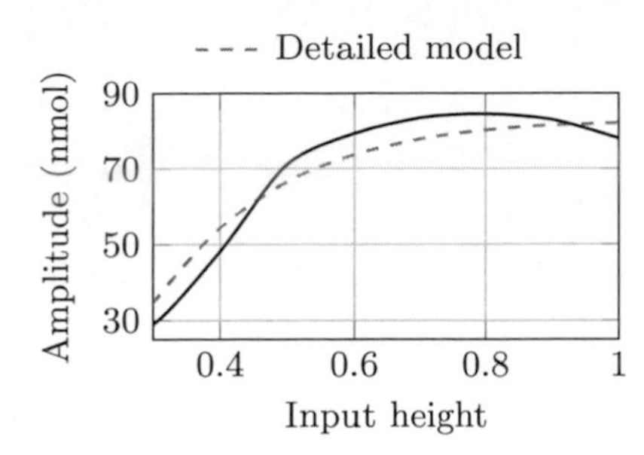

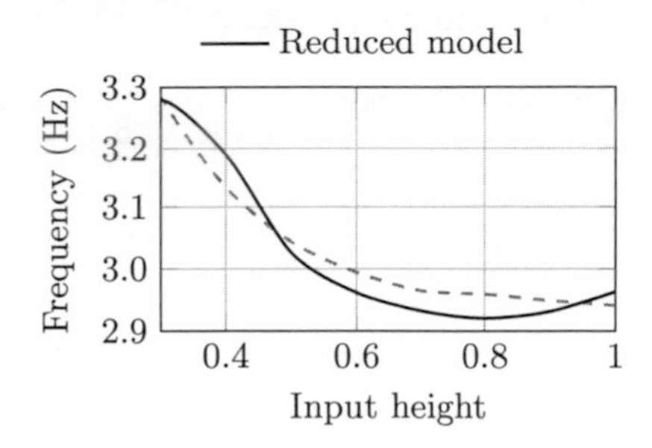

Figure 3.5: Amplitude and frequency of the limit cycle as a function of the input height.

can be detected and additionally the model can be reduced further while almost preserving the quantitative behavior for the considered set of inputs.

3.3 Preserving Stability in Trajectory-Based Model Reduction

In the previous section, we have addressed the first item of Problem 3.2. We have used Monte-Carlo integration to evaluate the objective functional, the observability normal form to parameterize the reduced model, and the equation error as introduced in Section 3.2.3 to end up with a convex optimization problem. In this section, we address the second item of Problem 3.2 and enforce the location and local exponential stability of the steady states in the set $\mathcal{S} = \left\{\left(u_{\mathrm{SS}}^{(s)}, y_{\mathrm{SS}}^{(s)}\right)\right\}_{s=1}^{n_{\mathrm{S}}}$ by adding constraints to the optimization problem. Hence, in the following we introduce a novel procedure to compute the parameters θ of the reduced model (3.3) such that the equation error is minimized while guaranteeing the location and local exponential stability of the steady states.

In Section 3.3.1, we derive conditions such that the location and exponential stability for a given set of steady states is ensured. These constraints are combined with the trajectory-based model reduction procedure in Section 3.3.2. The cone complementarity linearization is utilized in Section 3.3.3, to end up with a sequential convex optimization problem. The resulting model reduction algorithm is summarized in Section 3.3.4. In Section 3.3.5, the main results of applying the procedure to the Fermi-Pasta-Ulam lattice are presented.

3.3.1 Location and Stability of Steady States

In this section, we derive necessary and sufficient conditions for $\left(u_{\mathrm{SS}}^{(s)}, y_{\mathrm{SS}}^{(s)}\right)$ being a locally exponentially stable steady state of the reduced system.

Location of Steady States

In a first step, we ensure that $\left(u_{\mathrm{SS}}^{(s)}, y_{\mathrm{SS}}^{(s)}\right)$ is a steady state. Therefore, the right-hand side of the reduced model Σ_{R} has to vanish for $\xi = \begin{bmatrix} y_{\mathrm{SS}}^{(s)} & 0_{n_{\mathrm{R}}-1}^{\mathsf{T}} \end{bmatrix}^{\mathsf{T}}$ and $\mu = \begin{bmatrix} u_{\mathrm{SS}}^{(s)} & 0_{n_{\mathrm{R}}}^{\mathsf{T}} \end{bmatrix}^{\mathsf{T}}$. For $\dot{\xi}_1, \ldots, \dot{\xi}_{n_{\mathrm{R}}-1}$, this condition is fulfilled implicitly. Thus, the reduced model has a steady state at $\left(u_{\mathrm{SS}}^{(s)}, y_{\mathrm{SS}}^{(s)}\right)$ if and only if

$$0 = \theta^{\mathsf{T}} m\left(\begin{bmatrix} y_{\mathrm{SS}}^{(s)} \\ 0_{n_{\mathrm{R}}-1} \end{bmatrix}, \begin{bmatrix} u_{\mathrm{SS}}^{(s)} \\ 0_{n_{\mathrm{R}}} \end{bmatrix}\right) =: \theta^{\mathsf{T}} m^{(s)} .$$

Stability of Steady States

A necessary and sufficient condition for local exponential stability of $\left(u_{\mathrm{SS}}^{(s)}, y_{\mathrm{SS}}^{(s)}\right)$ is that all poles of the Jacobi linearization of Σ_{R} at $\left(u_{\mathrm{SS}}^{(s)}, y_{\mathrm{SS}}^{(s)}\right)$ lie in the open left half plane. Linearizing Σ_{R} at $\left(u_{\mathrm{SS}}^{(s)}, y_{\mathrm{SS}}^{(s)}\right)$ provides for the autonomous part

$$\Delta\dot{\xi} = \mathcal{A}^{(s)}(\theta)\Delta\xi \tag{3.6}$$

with $\mathcal{A}^{(s)}(\theta) = A + b\theta^{\mathsf{T}} C^{(s)}$ depending affinely on θ, and

$$A = \begin{bmatrix} 0_{n_{\mathrm{R}}-1} & I_{n_{\mathrm{R}}-1} \\ 0 & 0_{n_{\mathrm{R}}-1}^{\mathsf{T}} \end{bmatrix} \in \mathbb{R}^{n_{\mathrm{R}} \times n_{\mathrm{R}}},\ b = \begin{bmatrix} 0_{n_{\mathrm{R}}-1} \\ 1 \end{bmatrix} \in \mathbb{R}^{n_{\mathrm{R}}},\ C^{(s)} = \begin{bmatrix} c_1^{(s)} \\ \cdots \\ c_{n_{\mathrm{R}}}^{(s)} \end{bmatrix}^{\mathsf{T}} \in \mathbb{R}^{n_\theta \times n_{\mathrm{R}}},$$

$$c_k^{(s)} = \frac{\partial m}{\partial \xi_k}\left(\begin{bmatrix} y_{\mathrm{SS}}^{(s)} \\ 0_{n_{\mathrm{R}}-1} \end{bmatrix}, \begin{bmatrix} u_{\mathrm{SS}}^{(s)} \\ 0_{n_{\mathrm{R}}} \end{bmatrix}\right), \quad k = 1, \ldots, n_{\mathrm{R}} .$$

For a given θ the linear system (3.6) is exponentially stable if and only if there exists a symmetric matrix $X^{(s)} \in \mathbb{S}_{++}^{n_{\mathrm{R}}}$ satisfying the Lyapunov inequality [Boyd et al., 1994],

$$X^{(s)} \mathcal{A}^{(s)}(\theta) + \left(\mathcal{A}^{(s)}(\theta)\right)^{\mathsf{T}} X^{(s)} \prec 0, \qquad X^{(s)} \succ 0 . \tag{3.7}$$

To eliminate the multiplication of the two unknowns $X^{(s)}$ and θ, we use the following lemma, which is based on [Nguyen, 2004].

Lemma 3.7. *Given a constant matrix $\mathcal{A} \in \mathbb{R}^{n_{\mathrm{R}} \times n_{\mathrm{R}}}$ and symmetric matrix $X \in \mathbb{S}^{n_{\mathrm{R}}}$ with $X \succ 0$, the following statements are equivalent:*

1. $X\mathcal{A} + \mathcal{A}^{\mathsf{T}} X \prec 0$,
2. $\exists \rho \in \mathbb{R} : \begin{bmatrix} \rho X & -\mathcal{A}^{\mathsf{T}} - \rho I_{n_{\mathrm{R}}} \\ -\mathcal{A} - \rho I_{n_{\mathrm{R}}} & \rho X^{-1} \end{bmatrix} \succ 0 .$

Proof. Since $X \succ 0$, there exists a unique $Z = Z^\mathsf{T} \succ 0$ such that $Z^2 = X$. Hence, the square-root of X, denoted by $X^{1/2} := Z$, is well-defined. Therefore,

$$\begin{aligned} & X\mathcal{A} + \mathcal{A}^\mathsf{T} X \prec 0\,, \\ \Leftrightarrow \quad & X^{-1/2}\big(-X\mathcal{A} - \mathcal{A}^\mathsf{T} X\big)X^{-1/2} \succ 0\,, \\ \Leftrightarrow \quad & \begin{bmatrix} X^{-1/2} & X^{1/2} \end{bmatrix} \begin{bmatrix} 0_{n_\mathrm{R}\times n_\mathrm{R}} & -\mathcal{A}^\mathsf{T} \\ -\mathcal{A} & 0_{n_\mathrm{R}\times n_\mathrm{R}} \end{bmatrix} \begin{bmatrix} X^{-1/2} \\ X^{1/2} \end{bmatrix} \succ 0\,. \end{aligned} \tag{3.8}$$

Since the rows of $U := \begin{bmatrix} X^{1/2} & -X^{-1/2} \end{bmatrix}$ form a basis for the null space of $\begin{bmatrix} X^{-1/2} & X^{1/2} \end{bmatrix}$, Finsler's lemma [Boyd et al., 1994, page 33] provides that (3.8) is equivalent to the existence of $\rho \in \mathbb{R}$ such that

$$\begin{bmatrix} 0_{n_\mathrm{R}\times n_\mathrm{R}} & -\mathcal{A}^\mathsf{T} \\ -\mathcal{A} & 0_{n_\mathrm{R}\times n_\mathrm{R}} \end{bmatrix} + \rho U^\mathsf{T} U \succ 0\,, \tag{3.9}$$

which concludes the proof. □

Applying Lemma 3.7 to (3.7) yields the necessary and sufficient conditions for local exponential stability of $\big(u_\mathrm{SS}^{(s)}, y_\mathrm{SS}^{(s)}\big)$ given by

$$\begin{bmatrix} \rho^{(s)} X^{(s)} & -\big(\mathcal{A}^{(s)}(\theta)\big)^\mathsf{T} - \rho^{(s)} I_{n_\mathrm{R}} \\ -\mathcal{A}^{(s)}(\theta) - \rho^{(s)} I_{n_\mathrm{R}} & \rho^{(s)}\big(X^{(s)}\big)^{-1} \end{bmatrix} \succ 0\,, \qquad X^{(s)} \succ 0\,. \tag{3.10}$$

3.3.2 Model Reduction Ensuring Steady State Properties

From the above derivations follows the following theorem.

Theorem 3.8. *Consider the objective function $\hat{J}_m(\theta)$ given in (3.4), the set of steady states $\mathcal{S} = \big\{\big(u_\mathrm{SS}^{(s)}, y_\mathrm{SS}^{(s)}\big)\big\}_{s=1}^{n_\mathrm{S}}$, and the linearizations of Σ_R at the steady states $\mathcal{A}^{(s)}(\theta)$, $s = 1, \ldots, n_S$. A parameter vector θ ensures the location and local exponential stability of the steady states in $\mathcal{S}$, if and only if it is a feasible solution of*

$$\begin{aligned} \underset{\theta, X^{(s)}, \rho}{\text{minimize}} \quad & \hat{J}_m(\theta) \\ \text{subject to} \quad & \theta^\mathsf{T} m^{(s)} = 0\,, && \text{(C1)} \\ & X^{(s)} \succ 0\,, && \\ & \begin{bmatrix} \rho X^{(s)} & -\big(\mathcal{A}^{(s)}(\theta)\big)^\mathsf{T} - \rho I_{n_\mathrm{R}} \\ -\mathcal{A}^{(s)}(\theta) - \rho I_{n_\mathrm{R}} & \rho\big(X^{(s)}\big)^{-1} \end{bmatrix} \succ 0\,, && \text{(C2)} \\ & \text{for all } s = 1, \ldots, n_\mathrm{S}\,. \end{aligned}$$

The optimization of $\hat{J}_m(\theta)$ results in a minimization of the equation error.

In (C2), the same ρ is used for all steady states. This is possible, since from the positive definiteness of $U^{\mathsf{T}}U$ in (3.9) follows that (3.10) is also fulfilled when $\rho^{(s)}$ is replaced by any value that is larger. In consequence and to avoid the nonlinear terms $\rho X^{(s)}$ and $\rho\big(X^{(s)}\big)^{-1}$, we set ρ to a predefined large value.

Unfortunately, due to the appearance of $\big(X^{(s)}\big)^{-1}$ in (C2), the optimization problem in Theorem 3.8 is nonlinear and nonconvex. Therefore, in the following step a relaxation is performed to allow for the development of a computationally efficient algorithm.

3.3.3 Formulation as Sequential Convex Optimization

To overcome the nonconvexity, we propose a sequential convex programming approach using the cone complementarity linearization [El Ghaoui et al., 1997]. Therefore, $Y^{(s)} := \big(X^{(s)}\big)^{-1}$ is introduced into the optimization problem. Using $Y^{(s)}$, the constraint (C2) for the local exponential stability of a single steady state $\big(u_{\mathrm{SS}}^{(s)}, y_{\mathrm{SS}}^{(s)}\big)$ can be written as

$$X^{(s)} \succ 0\,, \tag{3.11}$$

$$Y^{(s)} \succ 0\,, \tag{3.12}$$

$$\begin{bmatrix} \rho X^{(s)} & -\big(\mathcal{A}^{(s)}(\theta)\big)^{\mathsf{T}} - \rho I_{n_{\mathrm{R}}} \\ -\mathcal{A}^{(s)}(\theta) - \rho I_{n_{\mathrm{R}}} & \rho Y^{(s)} \end{bmatrix} \succ 0\,, \tag{3.13}$$

$$X^{(s)}Y^{(s)} = I_{n_{\mathrm{R}}}\,. \tag{3.14}$$

Due to the equality constraint $X^{(s)}Y^{(s)} = I_{n_{\mathrm{R}}}$, this problem is still nonconvex. Nevertheless, an approximation of (3.14) can be derived by using the linear matrix inequality

$$\begin{bmatrix} X^{(s)} & I_{n_{\mathrm{R}}} \\ I_{n_{\mathrm{R}}} & Y^{(s)} \end{bmatrix} \succeq 0\,. \tag{3.15}$$

Therefore, note that (3.15) is feasible with $\operatorname{trace}\big(X^{(s)}Y^{(s)}\big) = n_{\mathrm{R}}$, $X^{(s)} \succ 0$, and $Y^{(s)} \succ 0$ if and only if $X^{(s)}Y^{(s)} = I_{n_{\mathrm{R}}}$ [El Ghaoui et al., 1997]. Otherwise $\operatorname{trace}\big(X^{(s)}Y^{(s)}\big) > n_{\mathrm{R}}$. Given this, a feasible solution for (3.11)–(3.14) may be found by solving the minimization problem

$$\begin{aligned} \underset{\theta, X^{(s)}, Y^{(s)}}{\text{minimize}} \quad & \operatorname{trace}\Big(X^{(s)}Y^{(s)}\Big) \\ \text{subject to} \quad & \text{(3.11)–(3.13) and (3.15).} \end{aligned}$$

Since for this optimization problem the objective function is nonlinear, a linear approximation is derived at a feasible point $\big(X_0^{(s)}, Y_0^{(s)}\big)$. This approximation is given by $\frac{1}{2}\operatorname{trace}\big(X_0^{(s)}Y^{(s)} + X^{(s)}Y_0^{(s)}\big)$. This yields the linearized problem

$$\begin{aligned} \underset{\theta, X^{(s)}, Y^{(s)}}{\text{minimize}} \quad & \frac{1}{2}\operatorname{trace}\Big(X_0^{(s)}Y^{(s)} + X^{(s)}Y_0^{(s)}\Big) \\ \text{subject to} \quad & \text{(3.11)–(3.13) and (3.15),} \end{aligned}$$

which is solved sequentially and shown to converge within a few iterations for similar problems [El Ghaoui et al., 1997].

Applying this procedure to the stability constraints (C2) in Theorem 3.8 and defining a mixed objective, we obtain

$$
\begin{aligned}
\underset{\theta, X^{(s)}, Y^{(s)}}{\text{minimize}} \quad & \hat{J}_m(\theta) + \alpha \sum_{s=1}^{n_\mathrm{S}} \operatorname{trace}\left(X_0^{(s)} Y^{(s)} + X^{(s)} Y_0^{(s)} \right) \\
\text{subject to} \quad & \left. \theta^\mathsf{T} m^{(s)} = 0 \,, \right\} \text{(C1)} \\
& \left. \begin{aligned}
& X^{(s)} \succ 0 \,, \\
& Y^{(s)} \succ 0 \,, \\
& \begin{bmatrix} \rho X^{(s)} & -\left(\mathcal{A}^{(s)}(\theta)\right)^\mathsf{T} - \rho I_{n_\mathrm{R}} \\ -\mathcal{A}^{(s)}(\theta) - \rho I_{n_\mathrm{R}} & \rho Y^{(s)} \end{bmatrix} \succ 0 \,, \\
& \begin{bmatrix} X^{(s)} & I_{n_\mathrm{R}} \\ I_{n_\mathrm{R}} & Y^{(s)} \end{bmatrix} \succeq 0 \,,
\end{aligned} \right\} \text{(RC2)} \\
& \text{for all } s = 1, \dots, n_\mathrm{S} \,,
\end{aligned}
$$

in which $\alpha \in \mathbb{R}_{++}$ determines the weighting for the multi-objective optimization. Instead of introducing an additional term into the objective function, one may also add the constraints

$$
\frac{1}{2} \operatorname{trace}\left(X_0^{(s)} Y^{(s)} + X^{(s)} Y_0^{(s)} \right) < (1 + \epsilon) n_\mathrm{R} \tag{3.16}
$$

in which $\epsilon \in \mathbb{R}_{++}$ and solve the following optimization problem.

Problem 3.9. *Convex optimization problem for stability preserving in trajectory-based model reduction*

$$
\left.
\begin{aligned}
\underset{\theta, X^{(s)}, Y^{(s)}}{\text{minimize}} \quad & \hat{J}_m(\theta) \\
\text{subject to} \quad & \text{(3.16), (C1), (RC2)} \,, \\
& \text{for all } s = 1, \dots, n_\mathrm{S}
\end{aligned}
\right\} \mathcal{OP}\left(\left\{ \left(X_0^{(s)}, Y_0^{(s)} \right) \right\}_{s=1}^{n_\mathrm{S}} \right)
$$

From our experience it is easier to determine a suitable ϵ than a weighting α.

3.3.4 Model Reduction Algorithm

Based on the above results, we propose Algorithm 3.10 below for model reduction ensuring steady state properties. For ease of presentation we omit the range of $s = 1, \dots, n_\mathrm{S}$ in Algorithm 3.10. The sequential convex optimization is stopped if the reduced model is exponentially stable in all steady states and if the objective criterion is less than γ.

Algorithm 3.10 (Model reduction ensuring steady state properties).
Require: $\{(X_0^{(s)}, Y_0^{(s)})\}$ *and* θ_0 *satisfying* (C1), (RC2), *and* (3.16)
$k \leftarrow 0$
while $\mathcal{A}^{(s)}(\theta_k)$ *not exponentially stable* ***or*** $\hat{J}_m(\theta_k) \geq \gamma$ ***do***
$\{(X_{k+1}^{(s)}, Y_{k+1}^{(s)})\}$, $\theta_{k+1} \leftarrow$ *Solve* $\mathcal{OP}(\{(X_k^{(s)}, Y_k^{(s)})\})$ *defined in Problem 3.9*
$k \leftarrow k+1$
end while
return θ_k

We set $X_0^{(s)} := I_{n_{\mathrm{R}}}$ and $Y_0^{(s)} := I_{n_{\mathrm{R}}}$ for $s = 1, \ldots, n_{\mathrm{S}}$ to initialize Algorithm 3.10, which was no restriction for the considered examples.

The complexity reduction by enforcing a sparse solution θ discussed in Section 3.2.4 can be applied also to Algorithm 3.10.

To summarize, we extend the trajectory-based model reduction procedure to ensure the location and local exponential stability of a predefined set of steady states. First, we derived necessary and sufficient conditions such that a tuple of an input and output value is a locally exponentially stable steady state of the reduced system. Second, we achieved a sequential convex optimization problem by utilizing the cone complementarity linearization.

3.3.5 Example: Fermi-Pasta-Ulam Lattice

In this section, the proposed model reduction approach is applied to the Fermi-Pasta-Ulam (FPU) lattice. The FPU lattice has been chosen due to its importance in nonlinear science and the existence of nonlinear oscillations. Due to the nonlinear oscillations, the poles of the Jacobi linearization around a steady state are close to or on the imaginary axis. Therefore, a reduced model from the procedure presented in Section 3.2 is prone to be not locally exponentially stable around the steady states.

The FPU model as in [Fermi et al., 1955, Figure 4] consisting of 16 particles with fixed end points and cubic forces is used. This results in a detailed model of order 28 with polynomials with 85 summands up to the third order. Since we consider an I/O based model reduction procedure, we define the input as an external force acting on particle 8 and the position of this particle as the output. The I/O trajectories are generated by 100 simulations in the time interval $[0, 100]$ using initial conditions around the steady state for an input of 1.5. The input trajectories start with 1.5 and consist of four smoothed steps varying the input in $[0.5, 2.5]$. Details and equations can be found in [Löhning et al., 2011b].

The set of steady states is defined by the two input values 0.75 and 2, which results in $\mathcal{S} = \{(0.75, 1.86), (2, 3.11)\}$. The ansatz functions of the reduced model are chosen to consist of all monomials with degree less or equal three. Several values for the order of the reduced model have been tested, yielding that a reasonable approximation of the I/O behavior is achieved for $n_{\mathrm{R}} = 2$.

Table 3.1: Poles of the Jacobi linearization at the steady states.

$u_{\mathrm{SS}}^{(s)}$	Reduced model with stability	Reduced model without stability
0.75	$-0.0147 \pm 0.336\jmath$	$0 \pm 0.328\jmath$
2.00	$-0.0125 \pm 0.475\jmath$	$0 \pm 0.482\jmath$

Applying Algorithm 3.10 with the sparsity enhancing ℓ_1-norm formulation of Section 3.2.4 to the FPU lattice results in the reduced model with local exponential stability with 13 parameters

$$\begin{aligned}\ddot{y}_{\mathrm{R}} =& -0.035\,y_{\mathrm{R}} - 0.04\,\dot{y}_{\mathrm{R}} + 0.14\,u_\varphi + 0.0019\,y_{\mathrm{R}}^2 - 0.0097\,y_{\mathrm{R}}\dot{y}_{\mathrm{R}} - 0.0001\,u_\varphi y_{\mathrm{R}} \\ &+ 0.011\,u_\varphi \dot{y}_{\mathrm{R}} - 0.011\,y_{\mathrm{R}}^3 + 0.017\,y_{\mathrm{R}}^2\dot{y}_{\mathrm{R}} + 0.011\,u_\varphi y_{\mathrm{R}}^2 - 0.035\,u_\varphi y_{\mathrm{R}}\dot{y}_{\mathrm{R}} \\ &- 0.0065\,u_\varphi^2 y_{\mathrm{R}} + 0.02\,u_\varphi^2 \dot{y}_{\mathrm{R}}\,.\end{aligned}$$

Hence, a reduction from the detailed model with 28 states and 85 summands to 2 states with 13 summands is achieved.

Furthermore, a second reduced model is computed to assess the influence of the constraint (RC2), which enforces the local exponential stability of the steady states. This second reduced model is determined by minimizing $\hat{J}_m(\theta)$ subject to the constraint (C1), which enforces only the location of the steady states. The optimization results in the reduced model without local exponential stability with 8 parameters

$$\begin{aligned}\ddot{y}_{\mathrm{R}} =& -0.038\,y_{\mathrm{R}} + 0.14\,u_\varphi + 0.057\,\ddot{u}_\varphi + 0.0095\,u_\varphi y_{\mathrm{R}} \\ &- 0.0074\,y_{\mathrm{R}}^3 - 0.076\,\dot{y}_{\mathrm{R}}^3 - 0.02\,u_\varphi^2 \dot{u}_\varphi + 0.96\,\dot{u}_\varphi^3\,.\end{aligned}$$

To validate the reduced models, simulations with new initial conditions and inputs are used. Exemplary trajectories are shown in Figure 3.6. The quantitative I/O behavior for increasing and decreasing steps is reproduced well by both reduced models. The second time derivative of the output demonstrates the major oscillation. The major oscillation is captured by the reduced models whereas the faster oscillations of the detailed model are not reproduced due to the limitation to two states.

To examine the local steady state behavior, the poles of the Jacobi linearization at the steady states of both reduced models are given in Table 3.1, showing that the second reduced model exhibits undamped oscillations. For the first reduced model the locally exponential stability is guaranteed by Algorithm 3.10 and results in lightly damped poles.

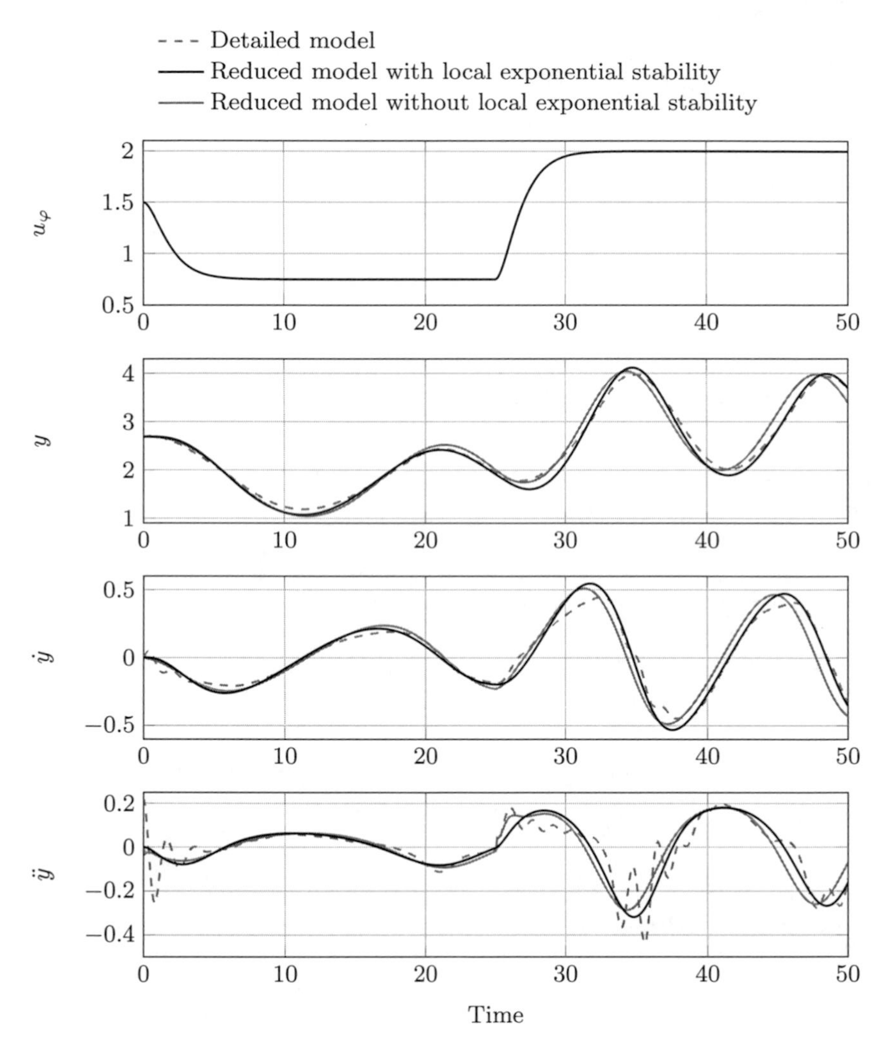

Figure 3.6: Trajectories of the detailed model of order 28 with 85 summands, the reduced model with local exponential stability of order 2 with 13 summands, and the reduced model without local exponential stability of order 2 with 8 summands.

3.4 Comparison with Approaches Relying Only on Simulated Trajectories

In the last sections the trajectory-based model reduction procedure was introduced. Before concluding this chapter, we compare the presented approach with existing approaches relying only on simulated trajectories cited in Section 1.2.1.

Lohmann [1994] presents a model reduction procedure that relies on the knowledge of the dominant states. Furthermore, the nonlinearity is kept in the reduced model and the linear couplings between the state variables and the nonlinear functions are computed. In contrast, the dominant states of the trajectory-based model reduction procedure are implicitly defined by the observability map and only the order of the reduced model has to be defined. Both methods have in common that the equation error is minimized. Model reduction of the MAPK cascade using the method of [Lohmann, 1994] is described in [Oppolzer, 2013]. A small error in the I/O behavior is achieved for a reduction to a model with three states for one fixed input trajectory. Unfortunately, according to Oppolzer [2013], the extension to a set of input trajectories resulted in a reduced model that does not reproduce the I/O behavior for inputs close to the border of the desired interval, whereas the trajectory-based approach lead to a reduced model that captures the quantitative I/O behavior of the MAPK cascade for inputs up to the limit of the desired interval as shown in Section 3.2.5.

In [Wood et al., 2004], derivatives of I/O data up to the second-order and an artificial neural network are used to determine an *implicit* nonlinear ODE. The approach presented in this thesis relies on the observability normal form, which results in an *explicit* nonlinear ODE.

Vargas and Allgöwer [2004] propose a procedure that is applicable to systems admitting a discrete-time Volterra representation and suggested an iterative approach for the construction of the reduced model. Volterra methods are known to generate local reduced models [Gu, 2011]. In comparison to [Vargas and Allgöwer, 2004], the trajectory-based approach uses the flexible system class described by the nonlinear observability normal form.

Another approach proposed in [Bond et al., 2010] uses a linear projection of the states. This reduced set of states is then used to optimize a continuous-time reduced model of the form $\dot{v} = f(u, v)$. In contrast, we use the observability normal form, which implicitly defines a nonlinear mapping from the states of the detailed model to the states of the reduced model. Bond et al. [2010] also showed that the approach allows to enforce incremental stability. The concept of incremental stability excludes multiple equilibria and is more strict than asymptotic stability, which can be enforced by the trajectory-based approach.

To summarize, the trajectory-based approach presented in this thesis uses the minimization of the equation error [Lohmann, 1994], the implicit definition of the states of the reduced model by the output and the derivatives thereof [Wood et al., 2004], and adding constraints to the optimization problem in order to enforce stability [Bond et al., 2010]. In contrast to the existing approaches, the trajectory-based approach uses the observability normal form for model reduction. The main advantage of the observability normal form is the implicitly defined nonlinear mapping from the states of the detailed model to the states of the reduced model.

3.5 Summary

We presented an I/O trajectory-based model reduction method for reducing the I/O map of continuous-time nonlinear ODEs. The method assumes that the detailed model evolves close to a — not necessarily flat — low-dimensional manifold. Furthermore, the reduced model is assumed to be globally observable. Thus, the nonlinear observability normal form can be used to parameterize the reduced model. Thereby, a nonlinear mapping from the states of the detailed model to the states of the reduced model is implicitly defined. Together with the possibility to define the relevant I/O behavior by means of the weighting function, the procedure typically results in reduced models of low order and computational complexity. The properties of the method are illustrated using a model of the MAPK cascade. We show that the model reduction procedure can cope with nonlinear behavior like limit cycles.

Furthermore, the trajectory-based nonlinear model reduction method is extended to preserve the location and local exponential stability of a set of steady states. The steady state constraints restrict the space of feasible parameters. To ensure a maximal number of degrees of freedom despite the stability requirement, necessary and sufficient conditions for simultaneous local exponential stability of the reduced model are derived. Since the resulting overall optimization problem has nonlinear constraints, the cone complementarity linearization is utilized. This results in a sequential convex optimization problem, which has good convergence properties. Hence, the model reduction for nonlinear ODEs could be reformulated to a series of efficient convex programs. To evaluate the proposed approach the FPU model is studied. Finally, the presented approach was compared with existing approaches relying only on simulated trajectories.

Chapter 4

A-Posteriori Bound for the Model Reduction Error

In the previous chapter, a method for model reduction of nonlinear systems has been presented. The model reduction introduces an error between the detailed model and the reduced model. In this chapter, we show how this error can be bounded while simulating the reduced model. Although the error bound introduced in this chapter can be used whenever a simulation with a reduced model is required, the main goal is to come up with an error bound, which can be utilized later in Chapters 5 and 6 for MPC. As discussed in Section 1.2.3 there are only few and limited results for MPC using reduced models that guarantee asymptotic stability or constraint satisfaction even for linear systems. To overcome these limitations, we also consider the class of LTI systems in this and the following chapters.

After stating the problem setup in Section 4.1 the error bounding system is derived in four steps. First, we introduce a preprocessing of the plant including a prestabilizing feedback and a state transformation in Section 4.2. Second, a bound for the norm of the matrix exponential is introduced in Section 4.3. This bound is a generalization of the bound utilized in the a-posteriori error bound presented in Section 2.2. Furthermore, we discuss in Section 4.3 the choice of the matrix defining the state transformation as well as the choice of the parameters of the generalized bound for the norm of the matrix exponential. Third, model reduction by projection is applied to the preprocessed model in Section 4.4. Fourth, the generalized bound for the norm of the matrix exponential is used to establish an a-posteriori error bound, which improves the result of Haasdonk and Ohlberger [2011] in terms of conservatism with comparable computational demand. The proposed error bound is then compared in Section 4.5 with existing ones using the model of the tubular reactor.

The main result of this chapter is visualized in Figure 4.1. The full order simulation is replaced by a reduced order simulation. The a-posteriori error bound establishes a relation between the state of the preprocessed model and the state of the reduced model. Furthermore, Figure 4.1 depicts the preprocessing of the plant.

This chapter is partially based on [Hasenauer, Löhning, Khammash, and Allgöwer, 2012; Löhning, Hasenauer, Khammash, and Allgöwer, 2011c] in which the improved error bound has already been published. The preprocessing of the plant and the application to the tubular reactor is based on [Löhning et al., 2014].

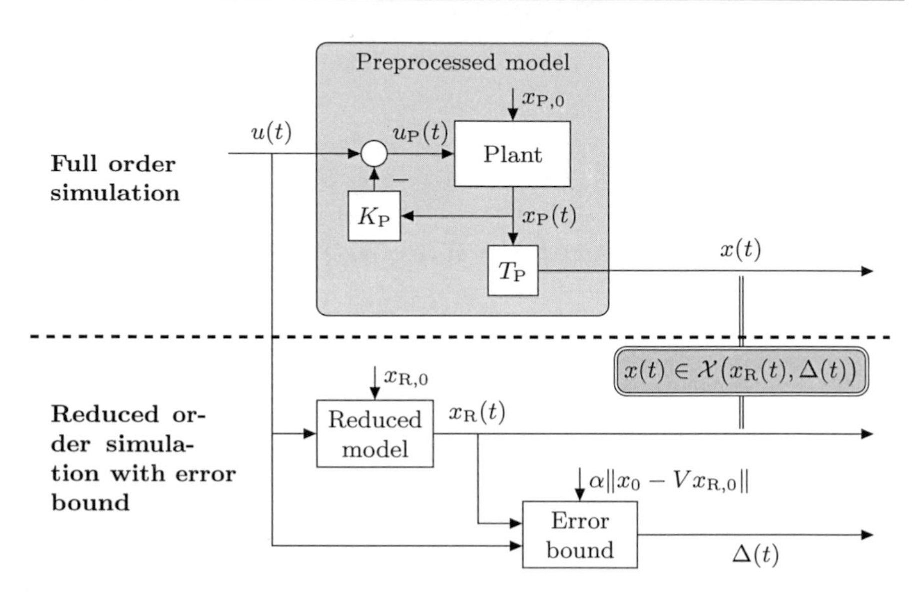

Figure 4.1: Structure of the simulation of the reduced model with the error bound and relation to the simulation of the preprocessed model. The connection between the preprocessed and the reduced model by the matrices V and W is omitted. The set of possible states of the preprocessed model $\mathcal{X}$ is defined in (4.9).

4.1 Problem Statement

In this chapter, we consider continuous-time LTI plants described by

$$\Sigma_P : \begin{cases} \dot{x}_P(t) = A_P x_P(t) + B_P u_P(t)\,, \\ x_P(0) = x_{P,0}\,, \end{cases} \tag{4.1}$$

in which $x_P(t) \in \mathbb{R}^n$ is the state vector at time t, $u_P(t) \in \mathbb{R}^{n_U}$ is the input vector at time t, and $u_P(\cdot) \in \mathcal{L}_2^{n_U}$. The model order n is assumed to be large, i.e., $n \gg 1$. In the following, we assume that all eigenvalues of A_P with nonnegative real part are controllable, which is ensured by the following assumption.

Assumption 4.1. *The pair (A_P, B_P) is stabilizable.*

Furthermore, we assume that the simulation of the plant model is computationally intractable for the application at hand due to the large model order. Examples are simulation studies for many inputs to check the proper operation or real-time requirements appearing in MPC. To increase the computational efficiency, a reduced model is used, which introduces an approximation error. To derive guarantees, we are interested in a bound for the model reduction error for all $t \geq 0$.

This bound can be used, e.g., in optimization tasks to minimize an upper bound for the performance criterion [Hasenauer et al., 2012] or in MPC to guarantee satisfaction of state constraints [Löhning et al., 2014].

The goal of this chapter is to improve the a-posteriori error bound introduced in Section 2.2 such that a tighter error bound with comparable computational demand is obtained. Furthermore, to guarantee asymptotic stability of MPC using reduced models in Chapter 5, we require that the improved error bound is given by an asymptotically stable system even for unstable plants.

An important assumption of the a-posteriori error bound presented in Theorem 2.1 in Section 2.2 is the constant bound α for the norm of the matrix exponential with a finite $\alpha \geq 1$, i.e.,

$$\left\| \mathrm{e}^{A_{\mathrm{P}} t} \right\| \leq \alpha \quad \text{for all } t \geq 0 . \tag{4.2}$$

In the following two sections, we address issues arising from this assumption.

4.2 Preprocessing of the Plant

A necessary condition for the bound (4.2) is that the eigenvalue with the largest real part $\lambda_{\max}(A_{\mathrm{P}})$ is less or equal to zero. Hence, the error bound in Theorem 2.1 can only be applied to stable LTI systems. To extend the error bound to unstable systems, it is in general possible to limit the error bound to a finite time interval [Haasdonk and Ohlberger, 2011]. Alternatively, the unstable modes can be decoupled and retained in the reduced model as discussed below in Remark 4.5. In [Löhning et al., 2014], the authors exploit that the input is designed in MPC and the state of the plant is assumed to be known. This makes a feedback of the model reduction error possible, which allows to stabilize the error dynamics.

Motivated by the application of the error bound for MPC in Chapter 5, we allow for a prestabilization of the plant

$$u_{\mathrm{P}}(t) := u(t) - K_{\mathrm{P}} x_{\mathrm{P}}(t) ,$$

which is an approach familiar in MPC [Kouvaritakis et al., 2000; Mayne et al., 2005]. Thus, we assume that the full state can be measured. This assumption can be relaxed if the feedback requires only a subspace of the state space as discussed later in Proposition 5.10. We emphasize that for asymptotically stable systems the prestabilization is not necessary for the improvement of the error bound discussed in Section 4.3. Hence, the improved error bound is of interest in a far broader context than used in Chapter 5.

In general, using a suitable norm to bound the model reduction error and the norm of the matrix exponential is less conservative. Haasdonk and Ohlberger [2011] use a weighted vector norm $\|x\|_G := \sqrt{x^{\mathsf{T}} G x}$ for some $G \succ 0$ and the matrix norm induced by the weighted vector norm. Using a weighted norm would make the derivations in Chapter 5 more complicated. Hence, we alternatively apply a linear state transformation to the prestabilized plant

$$x(t) := T_{\mathrm{P}} x_{\mathrm{P}}(t),$$

in which $T_{\mathrm{P}} \succ 0$. How to compute T_{P} is discussed in the following section.

The combination of the prestabilization and state transformation is denoted preprocessing. This results in the preprocessed model as visualized in Figure 4.1. The preprocessed model is given by

$$\Sigma_{\mathrm{PP}} : \begin{cases} \dot{x}(t) = Ax(t) + Bu(t), \\ x(0) = x_0 \end{cases} \tag{4.3}$$

with $A = T_{\mathrm{P}}(A_{\mathrm{P}} - B_{\mathrm{P}}K_{\mathrm{P}})T_{\mathrm{P}}^{-1}$, $B = T_{\mathrm{P}}B_{\mathrm{P}}$, and $x_0 = T_{\mathrm{P}}x_{\mathrm{P},0}$.

To apply the bound for the error between the detailed and reduced model stated in Theorem 2.1 to the preprocessed model, the model reduction by projection has to be applied to the preprocessed model. Note that the detailed model is chosen as the preprocessed model (see Figure 2.3 on page 23).

The advantage of the preprocessing is that the bound for the norm of the matrix exponential (4.2) has to be satisfied for A instead of the plant dynamics A_P. Since the pair $(A_{\mathrm{P}}, B_{\mathrm{P}})$ is stabilizable, there exists a K_{P} such that A is Hurwitz. Hence, a finite and constant bound α for the norm of the matrix exponential exists and the error bound in Theorem 2.1 can be applied to the preprocessed model.

Löhning et al. [2014] use a linear feedback of the model reduction error, i.e., $u_{\mathrm{P}}(t) = u(t) - K_{\mathrm{P}}\big(x_{\mathrm{P}}(t) - Vx_{\mathrm{R}}(t)\big)$ instead of the prestabilization. The advantage of the preprocessing is that the prestabilization and state transformation is included in the preprocessed model. Thereby, the preprocessed model serves as an interface to the error bound and subsequent steps like MPC as visualized in Figure 4.1. Consequently, the preprocessing does not appear directly in the error bound and the model predictive controller. Altogether, we provide a modular approach that hides the complexity of the module preprocessing from the error bound and subsequent modules like the MPC approach discussed in the following chapters. This modularity simplifies the notation as well as the derivations especially in Chapter 5 and 6.

Summarizing, the prestabilization allows to apply the error bound of Theorem 2.1 to unstable plants. Furthermore, the state transformation reduces the conservatism by implicitly introducing a suitable norm.

4.3 A Bound for the Norm of the Matrix Exponential

The constant bound for the norm of the matrix exponential in (4.2) results in an error bound that is a monotonically increasing function of time as observable in (2.5). To achieve a tighter error bound, we generalize (4.2) by introducing an exponential decay rate.

Assumption 4.2. *The norm of the matrix exponential for the preprocessed model (4.3) is bounded by*

$$\left\| \mathrm{e}^{At} \right\| \leq \alpha \, \mathrm{e}^{-\beta t} \quad \textit{for all } t \geq 0$$

with $\alpha \geq 1$ *and* $\beta \in \mathbb{R}$.

For the parameters α and β in Assumption 4.2, there is in general a trade-off between a small $\alpha \geq 1$ and a large exponential decay rate β. If we choose $\alpha = 1$, then the exponential decay rate is given by $\beta = -0.5\lambda_{\max}(A + A^{\mathsf{T}})$ [Hinrichsen and Pritchard, 2005, Lemma 5.5.11], which can be smaller than zero even if A is Hurwitz. Alternatively, one can choose $\beta < -\lambda_{\max}(A)$ and any $\tilde{Q} \in \mathbb{S}^n_{0+}$. Then, $\alpha = \left(\sigma_{\max}(\tilde{T}_{\mathrm{P}})/\sigma_{\min}(\tilde{T}_{\mathrm{P}})\right)^{1/2}$ is determined by the solution $\tilde{T}_{\mathrm{P}} \in \mathbb{S}^n_{++}$ of the Lyapunov equation

$$(A + \beta I_n)^{\mathsf{T}} \tilde{T}_{\mathrm{P}} + \tilde{T}_{\mathrm{P}}(A + \beta I_n) + \tilde{Q} = 0\,, \tag{4.4}$$

see [Hinrichsen and Pritchard, 2005, page 664f].

Furthermore, the matrix A depends on the preprocessing of the plant and, hence, on the choice of K_{P} and T_{P}. Consider that K_{P} is chosen such that $A_{\mathrm{P}} - B_{\mathrm{P}}K_{\mathrm{P}}$ is Hurwitz. We use the degree of freedom given by the transformation matrix T_{P} in order to satisfy Assumption 4.2 with $\alpha = 1$ and a decay rate $0 < \beta < -\lambda_{\max}(A_{\mathrm{P}} - B_{\mathrm{P}}K_{\mathrm{P}})$. To this end, we propose to solve the following optimization problem to obtain the state transformation matrix $T_{\mathrm{P}} = \tilde{T}_{\mathrm{P}}^{1/2} \in \mathbb{S}^n_{++}$ (compare [Hinrichsen et al., 2002, (4.1)]).

Problem 4.3 (Computation of the state transformation matrix)**.**

$$\begin{aligned}
&\underset{\gamma, \tilde{T}_{\mathrm{P}} \in \mathbb{S}^n_{++}}{\text{minimize}} && \gamma \\
&\text{subject to} && (A_{\mathrm{P}} - B_{\mathrm{P}}K_{\mathrm{P}})^{\mathsf{T}} \tilde{T}_{\mathrm{P}} + \tilde{T}_{\mathrm{P}}(A_{\mathrm{P}} - B_{\mathrm{P}}K_{\mathrm{P}}) + 2\beta \tilde{T}_{\mathrm{P}} \preceq 0\,, && \text{(4.5a)} \\
& && I_n \preceq \tilde{T}_{\mathrm{P}} \preceq \gamma I_n\,. && \text{(4.5b)}
\end{aligned}$$

Problem 4.3 is a semidefinite program [Boyd and Vandenberghe, 2004, page 168] and can be solved efficiently. Since $A_{\mathrm{P}} - B_{\mathrm{P}}K_{\mathrm{P}} + \beta I_n$ is Hurwitz, the existence of a solution is guaranteed by properties of the Lyapunov equation [Bernstein, 2009, Proposition 11.9.5].

Inserting $\tilde{T}_{\mathrm{P}} := T_{\mathrm{P}}^2$ into (4.5a), multiplying with T_{P}^{-1} from left and right, and using $A = T_{\mathrm{P}}(A_{\mathrm{P}} - B_{\mathrm{P}}K_{\mathrm{P}})T_{\mathrm{P}}^{-1}$ results in $A^{\mathsf{T}} + A + 2\beta I_n \preceq 0$. Hence, (4.5a) implies that Assumption 4.2 is satisfied for $\alpha = 1$ [Hinrichsen and Pritchard, 2005, Corollary 5.5.26].

The constraints in (4.5b) enforce $\lambda_{\min}(T_{\mathrm{P}}) \geq 1$ and $\lambda_{\max}(T_{\mathrm{P}}) \leq \gamma^2$. Hence, the objective minimizes the condition number

$$\kappa(T_{\mathrm{P}}) := \left\|T_{\mathrm{P}}\right\| \left\|T_{\mathrm{P}}^{-1}\right\| = \sigma_{\max}(T_{\mathrm{P}})/\sigma_{\min}(T_{\mathrm{P}}) = \lambda_{\max}(T_{\mathrm{P}})/\lambda_{\min}(T_{\mathrm{P}})\,,$$

which measures the deformation of the state transformation [Hinrichsen et al., 2002].

If no state transformation is desired, i.e., $T_\mathrm{P} = I_n$, then we have $A = A_\mathrm{P} - B_\mathrm{P} K_\mathrm{P}$. Consider $\tilde{Q} := -(A + \beta I_n)^\mathsf{T} \tilde{T}_\mathrm{P} - \tilde{T}_\mathrm{P}(A + \beta I_n)$. Then, from (4.5a) follows $\tilde{Q} \in \mathbb{S}^n_{0+}$. Furthermore, for this choice of $\tilde{Q}$, (4.4) is satisfied. Hence, if no state transformation is desired, the matrix $\tilde{T}_\mathrm{P}$ can be used to compute $\alpha = \big(\sigma_{\max}(\tilde{T}_\mathrm{P})/\sigma_{\min}(\tilde{T}_\mathrm{P})\big)^{1/2}$.

In this section, we have introduced the generalized bound for the norm of the matrix exponential. Furthermore, we have discussed how the parameters α and β appearing in the bound as well as the state transformation matrix T_P can be computed. In the following, we assume that the parameters K_P, T_P, $\alpha \geq 1$, and $\beta \in \mathbb{R}$ are known. Furthermore, we only consider the preprocessed model and not the plant model.

4.4 Asymptotically Stable Error Bounding System

In this section, we use the generalized bound for the norm of the matrix exponential given by Assumption 4.2 to improve the error bound of Theorem 2.1. Furthermore, we discuss two ways to achieve an asymptotically stable error bounding system and the relation to existing a-posteriori error bounds.

4.4.1 Improved A-Posteriori Error Bound

As for the error bound given by Theorem 2.1, the result of this section applies to any model reduction method based on a projection as introduced in Section 2.1.1. Consequently, many of the linear model reduction methods may be used, e.g., most of the methods described in [Antoulas, 2005b]. Therefore, we assume that the matrices $V \in \mathbb{R}^{n \times n_\mathrm{R}}$ and $W \in \mathbb{R}^{n \times n_\mathrm{R}}$ satisfying $W^\mathsf{T} V = I_{n_\mathrm{R}}$ are given and result in the reduced model

$$\Sigma_\mathrm{R} : \begin{cases} \dot{x}_\mathrm{R}(t) = A_\mathrm{R} x_\mathrm{R}(t) + B_\mathrm{R} u(t)\,, \\ x_\mathrm{R}(0) = x_{\mathrm{R},0}\,, \end{cases} \tag{4.6}$$

in which $A_\mathrm{R} = W^\mathsf{T} A V$ and $B_\mathrm{R} = W^\mathsf{T} B$.

For the setup at hand, we have the following bound for the model reduction error.

Theorem 4.4 (Improved a-posteriori error bound)**.** *Consider the preprocessed model* (4.3)*, the reduced model* (4.6) *defined by* V, W *and that Assumption 4.2 is satisfied by* $\alpha \geq 1$ *and* $\beta \in \mathbb{R}$*. Then, the model reduction error is bounded by*

$$\|x(t) - V x_\mathrm{R}(t)\| \leq \Delta(t) \quad \text{for all } t \geq 0\,,$$

in which $\Delta(t)$ *is defined by the error bounding system*

$$\Sigma_\Delta : \begin{cases} \dot{\Delta}(t) = -\beta \Delta(t) + \alpha \|r(t)\|\,, \\ \Delta(0) = \alpha \|x(0) - V x_\mathrm{R}(0)\|\,, \end{cases} \tag{4.7}$$

with the residual $r(t) = \big(I_n - V W^\mathsf{T}\big)\big(A V x_\mathrm{R}(t) + B u(t)\big)$.

Proof. The proof is based on [Haasdonk and Ohlberger, 2011, Proof of Proposition 1], which has to be extended by the exponential decay rate.

To derive the error bound, consider the error

$$e(t) := x(t) - Vx_{\mathrm{R}}(t)$$

between the state of the preprocessed model $x(t)$ and the estimated state $Vx_{\mathrm{R}}(t)$. The error dynamics are given by

$$\begin{aligned}\dot{e}(t) &= \dot{x}(t) - V\dot{x}_{\mathrm{R}}(t) \\ &= Ax(t) + Bu(t) - VA_{\mathrm{R}}x_{\mathrm{R}}(t) - VB_{\mathrm{R}}u(t) \\ &= Ae(t) + \big(AV - VA_{\mathrm{R}}\big)x_{\mathrm{R}}(t) + \big(B - VB_{\mathrm{R}}\big)u(t) \\ &= Ae(t) + r(t)\,.\end{aligned}$$

The error dynamics have the solution

$$e(t) = \mathrm{e}^{At}\, e(0) + \int_0^t \mathrm{e}^{A(t-\tau)}\, r(\tau)\,\mathrm{d}\tau\,.$$

By applying the norm and using the sub-multiplicative property of the induced matrix norm we obtain

$$\|e(t)\| \leq \left\|\mathrm{e}^{At}\right\| \|e(0)\| + \int_0^t \left\|\mathrm{e}^{A(t-\tau)}\right\| \|r(\tau)\|\,\mathrm{d}\tau\,.$$

Using Assumption 4.2, the error is bounded from above by

$$\|e(t)\| \leq \Delta(t) = \alpha\,\mathrm{e}^{-\beta t}\, \|e(0)\| + \alpha \int_0^t \mathrm{e}^{-\beta(t-\tau)}\, \|r(\tau)\|\,\mathrm{d}\tau\,. \tag{4.8}$$

To conclude the proof, we observe that the error bound $\Delta(t)$ is the solution of the error bounding system (4.7). □

The error bound derived in Theorem 4.4 gives a relation between the state of the preprocessed model and the state of the reduced model. More precisely, the set of possible states of the preprocessed model is

$$\mathcal{X}\big(x_{\mathrm{R}}(t), \Delta(t)\big) := \{x \in \mathbb{R}^n \mid \|x - Vx_{\mathrm{R}}(t)\| \leq \Delta(t)\}\,. \tag{4.9}$$

This set relates the simulation of the reduced model with the simulation of the preprocessed model as indicated in Figure 4.1 on page 60. Furthermore, Figure 4.1 visualizes the cascaded structure of the reduced model with the error bound. We emphasize that only $u(t)$, $x_{\mathrm{R}}(t)$, and $\|x_0 - Vx_{\mathrm{R},0}\|$ are required to compute the error bound.

The time dependent bound $\|\mathrm{e}^{At}\| \leq \alpha\,\mathrm{e}^{-\beta t}$ utilized in Theorem 4.4 enables, for $\beta > 0$, an exponentially decaying influence of the initial error and history of the residual. Compared to the bound $\|\mathrm{e}^{At}\| \leq \alpha$ used in [Haasdonk and Ohlberger, 2011], the exponential decay in general results in tighter error bounds especially over a long time horizon. Moreover, the exponential decay by $\beta > 0$ is utilized to prove asymptotic stability of the MPC schemes in the following chapters.

4.4.2 Achieving an Asymptotically Stable Error Bounding System

The condition $\beta > 0$ implies that the preprocessed model needs to be asymptotically stable. This can be attained by the prestabilizing feedback introduced in Section 4.2. Instead of the prestabilization, an asymptotically stable error bounding system for unstable preprocessed models can also be achieved by limiting the model reduction method. The following remark is inspired by [Dubljevic et al., 2006], in which the dominant states are decoupled and a different error bound for the nondominant states is derived.

Remark 4.5 (Error bound for decoupled dominant states). *Consider a possibly unstable preprocessed model and model reduction by projection as described in Section 2.1.1 with a nonsingular matrix T defining the state transformation $x(t) = Tz(t)$ such that the dominant states depend only on the input and, in particular, not on the nondominant states,*

$$\begin{bmatrix} \dot{z}_1(t) \\ \dot{z}_2(t) \end{bmatrix} := \begin{bmatrix} A_1 & 0_{n_\mathrm{R} \times n - n_\mathrm{R}} \\ A_{21} & A_2 \end{bmatrix} \begin{bmatrix} z_1(t) \\ z_2(t) \end{bmatrix} + \begin{bmatrix} B_1 \\ B_2 \end{bmatrix} u(t) \,, \quad \begin{bmatrix} z_1(0) \\ z_2(0) \end{bmatrix} = T^{-1} x_0 = \begin{bmatrix} W^\mathrm{T} \\ \tilde{W}^\mathrm{T} \end{bmatrix} x_0 \,. \tag{4.10}$$

We allow for a lower left block A_{21} unequal zero since this allows numerically better conditioned matrices T. Truncation of $z_2(t)$ results in the reduced model

$$\dot{x}_\mathrm{R}(t) = A_1 x_\mathrm{R}(t) + B_1 u(t) \,, \quad x_\mathrm{R}(0) = z_1(0) \,.$$

Since $\dot{z}_1(t) = \dot{x}_\mathrm{R}(t)$, the error satisfies

$$e_z(t) := z(t) - \begin{bmatrix} x_\mathrm{R}(t) \\ 0_{n-n_\mathrm{R}} \end{bmatrix} = \begin{bmatrix} z_1(t) - x_\mathrm{R}(t) \\ z_2(t) \end{bmatrix} = \begin{bmatrix} 0_{n_\mathrm{R}} \\ z_2(t) \end{bmatrix} .$$

Hence, only an error bound for the nondominant states $z_2(t)$ is required. Consequently, the dynamics of the nondominant states z_2 have to be bounded, which are typically asymptotically stable. If $\|\mathrm{e}^{A_2 t}\| \leq \alpha_2 \, \mathrm{e}^{-\beta_2 t}$ for all $t \geq 0$, then the following error bound can be derived similar to the proof of Theorem 4.4.

$$\|z_2(t)\| \leq \alpha_2 \, \mathrm{e}^{-\beta_2 t} \, \|z_2(0)\| + \alpha_2 \int_0^t \mathrm{e}^{-\beta_2 (t-\tau)} \, \|A_{21} z_1(\tau) + B_2 u(\tau)\| \, \mathrm{d}\tau \,.$$

A well-known model reduction method that decouples the dynamics of the dominant states from the nondominant states is modal truncation. In a nutshell, modal truncation performs an eigenvalue decomposition of A, i.e., $AT = \Lambda T$. Assuming that A is diagonalizable, then Λ is a diagonal matrix and the eigenvalues appearing on the diagonal of Λ can be ordered according to a dominance measure, e.g., with decreasing real part. Hence, the reduced model preserves the n_R dominant poles. For details, we refer to [Antoulas, 2005a; Varga, 1995] and the references therein. For some systems modal truncation may be appropriate, e.g., systems with a small number of unstable and dominant modes as the example considered in [Dubljevic et al., 2006]. However, modal truncation may result in a bad approximation if the eigenvalue decomposition contains Jordan blocks [Antoulas, 2005a].

Summarizing, the error dynamics can be changed by a prestabilization or by decoupling some modes and bounding only the error of the modes affected by the model reduction. The combination of prestabilization followed by decoupling some modes is also possible.

4.4.3 Relation to Existing A-Posteriori Error Bounds

The deduced error bound can be directly extended to the class of linear time varying systems with an affine parameter dependence of the system matrices. Hence, the error bound is a generalization of the error bound introduced in [Haasdonk and Ohlberger, 2011], which is recovered for $\alpha = \sup_{t\geq 0} \left\|e^{At}\right\|$ and $\beta = 0$. In contrast to [Ruiner et al., 2012], where the norm of the matrix exponential is bounded by a general time dependent function, our error bound can still be implemented by a scalar ODE. Hence, the computational complexity of the proposed error bound is much lower than the generalization discussed in [Ruiner et al., 2012]. The statements concerning the computational efficiency of the error bound in Section 2.2 apply also to the improved error bound in Theorem 4.4.

For MPC as considered in the subsequent chapters, the error bounding system appears in the online optimization problem. Since this increases the computational complexity, constant bounds for the model reduction error are often used, e.g., [Dubljevic et al., 2006; Kögel and Findeisen, 2015; Lorenzetti et al., 2019]. A constant bound can be derived from the proposed error bound if the states and inputs of the reduced model are constrained to a compact set.

Remark 4.6 (Constant error bound for worst-case states and inputs). *Let the states and inputs of the reduced model be constrained to a compact set $\mathcal{C}$. Then, the norm of the residual is bounded by*

$$c_{\mathrm{r}} := \max_{(x_{\mathrm{R}}(t),u(t))\in\mathcal{C}} \left\| \left(I_n - VW^{\mathsf{T}}\right)\left(AVx_{\mathrm{R}}(t) + Bu(t)\right)\right\| .$$

Hence, from (4.8) *follows for* $\beta > 0$

$$\begin{aligned} \|x(t) - Vx_{\mathrm{R}}(t)\| &\leq \alpha\, e^{-\beta t}\, \|x(0) - Vx_{\mathrm{R}}(0)\| + \alpha c_{\mathrm{r}} \int_0^t e^{-\beta(t-\tau)}\, \mathrm{d}\tau \\ &\leq \alpha \|x(0) - Vx_{\mathrm{R}}(0)\| + \alpha c_{\mathrm{r}}/\beta\,. \end{aligned} \tag{4.11}$$

Since the worst-case states and inputs of the reduced model instead of the known values are used, the constant error bound in (4.11) *can be significantly more conservative than the error bound proposed in Theorem 4.4.*

Remark 4.5 and Remark 4.6 can be combined. This results in a constant error bound for the nondominant states. This combination is of interest within this thesis since a similar error bound is utilized in [Dubljevic et al., 2006] for MPC using reduced models in order to guarantee constraint satisfaction. Compared to [Dubljevic et al., 2006] in this thesis the plant models are described by ODEs, instead of linear parabolic PDEs, and the model reduction is not limited to modal truncation.

Summarizing, the generalized bound for the norm of the matrix exponential is used to deduce an improved error bound. This error bound is a generalization of the error bound in [Haasdonk and Ohlberger, 2011]. The advantage of the generalized error bound over the error bounds in [Dubljevic et al., 2006; Haasdonk and Ohlberger, 2011] is demonstrated by the tubular reactor in the following section. The comparison with the error bound of Remark 4.6 shows that taking the actual input and states into account leads to a significantly tighter error bound for the tubular reactor.

4.5 Example: Tubular Reactor

To evaluate the improved bound for the model reduction error, the tubular reactor introduced in Section 2.4 is considered in this section. At the beginning, the influence of the preprocessing on the bound for the norm of the matrix exponential is studied. Afterwards, the preprocessed plant model is reduced. Then, the proposed bound for the model reduction error is compared with the error bound presented in [Haasdonk and Ohlberger, 2011]. Finally, the effect of the model reduction and the error bound on the computation time is assessed.

4.5.1 Preprocessing and Bounding the Norm of the Matrix Exponential

The conservatism of the bound for the model reduction error introduced in Theorem 4.4 is significantly influenced by the parameters α and β, which define the bound for the norm of the matrix exponential. Hence, we study in the following how the bound for the norm of the matrix exponential changes with the prestabilization and state transformation introduced in Section 4.2.

The linear plant model of the tubular reactor

$$\frac{\mathrm{d}\Delta x_{\mathrm{P}}(t)}{\mathrm{d}t} = A_{\mathrm{P}}\,\Delta x_{\mathrm{P}}(t) + B_{\mathrm{P}} u_{\mathrm{P}}(t)\,, \qquad \Delta x_{\mathrm{P}}(0) = x_{\mathrm{SP}}(T_{\mathrm{nom}}) - x_{\mathrm{SP}}(T_{\mathrm{in}})\,,$$

see also (2.18) on page 35 and (A.2) on page 131, consists of three separate integrators with an asymptotically stable system in series. Hence the norm of the matrix exponential of A_{P} converges for $t \to \infty$ to its supremum, which is 34.3. We prestabilize the plant model with the linear-quadratic regulator (LQR). After prestabilization the supremum of the norm of the matrix exponential is 5.7. Hence, it is reduced by more than a factor of 6.

A state transformation that satisfies Assumption 4.2 for $\alpha = 1$ is used in the following subsections and chapters since it results in a smaller error bound. The

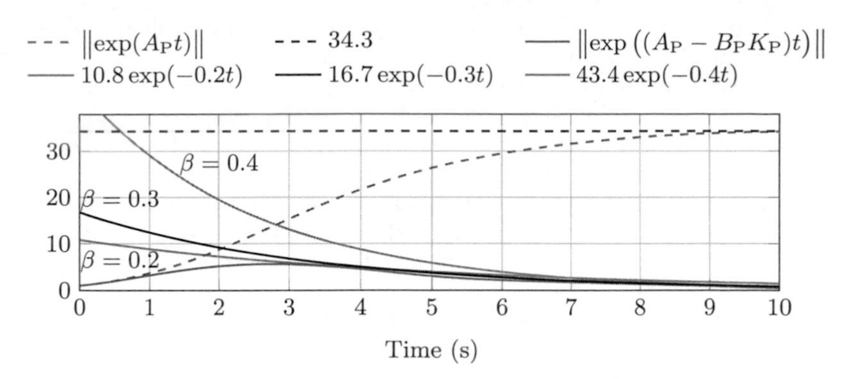

Figure 4.2: Comparison of the norm of the matrix exponential with upper bounds for the tubular reactor without prestabilization (dashed) and with prestabilization (solid).

state transformation is computed according to Problem 4.3, which minimizes the condition number of the state transformation matrix. Problem 4.3 is solved using YALMIP [Löfberg, 2009] with the solver SDPT3 [Toh et al., 1999]. Upper bounds for the norm of the matrix exponential of the prestabilized plant for different values of the decay rate β are visualized in Figure 4.2. For each β, we use a different state transformation that results in $\alpha = 1$. The bounds are scaled with the condition number of the state transformation matrix in Figure 4.2 to incorporate the deformation of the state transformation. Thereby, Figure 4.2 allows to judge the conservatism of the different bounds for the norm of the matrix exponential.

For $\beta = 0$, the condition number of the state transformation matrix is equal to the supremum of the norm of the matrix exponential, which is 5.7. The decay rate β can be increased up to $-\lambda_{\max}(A_P - B_P K_P) = 0.51$, which results in a state transformation with larger condition number as discussed in Section 4.3. In the following, we use the compromise between a small condition number, i.e., deformation of the state transformation, and a large decay rate given by $\beta = 0.3$ and $\kappa(T_P) = 16.7$.

Summarizing, the prestabilization results in a better decay rate and smaller condition number of the state transformation required to achieve $\alpha = 1$.

4.5.2 Model Reduction

The reduced model is derived in the style of [Agudelo et al., 2007a,b] and relies on the POD as introduced in Section 2.1.2. Since the model reduction by projection has to be applied to the preprocessed plant as discussed in Section 4.2, we use the preprocessed model (4.3) to compute the snapshot for the POD. In the following chapter, the reduced model is used for MPC. Thus, the set of initial conditions and input trajectories used to compute the snapshot should represent the trajectories occurring in closed loop [Marquardt, 2002, page 36]. As we will see later in Section 5.5, the model predictive controller deviates the input of the plant from the LQR solution in order to fulfill the input and state constraints. Since the preprocessed plant model is prestabilized by the LQR, representative closed-loop trajectories can be generated by exciting the preprocessed model by T_{in} and small deviations from the LQR solution $u(t)$. The inlet fluid temperature T_{in} leads to an excitation of the initial condition due to the shift of the setpoint as discussed in Section 2.4.2. For the inlet fluid temperature 20 values within $[315\,\mathrm{K}, 365\,\mathrm{K}]$ have been taken. The deviation of the normalized time derivatives of the jacket temperatures from the LQR solution $u(t)$ are chosen as piecewise constant. The input changes at $t = 0\,\mathrm{s}$, $3.3\,\mathrm{s}$, and $6.7\,\mathrm{s}$. The step height is varied for every inlet fluid temperature and taken from a uniform distribution in $[-0.05\,T_{\mathrm{nom}}\,{}^{1}\!/\mathrm{s}, 0.05\,T_{\mathrm{nom}}\,{}^{1}\!/\mathrm{s}]$. The time responses of the preprocessed model in the interval $t \in [0\,\mathrm{s}, 10\,\mathrm{s}]$ are sampled every $0.05\,\mathrm{s}$ as in [Agudelo et al., 2007a] to obtain the snapshot.

To compute the singular vectors of the snapshot matrix the implementation contained in RBmatlab [Drohmann et al., 2012a] is used, since it resulted in a lower error bound compared to the SVD of MATLAB. To have feasibility of the open loop optimal control problem occurring later for MPC, we choose $n_{\mathrm{R}} = 40$. Finally, the reduced model is computed via projection.

A validation of the reduced model is implicitly done in the following subsection by looking at the bound of the model reduction error.

4.5.3 A-Posteriori Bound for the Model Reduction Error

Given the reduced model and the bound for the norm of the matrix exponential, the error bounding system (4.7) can be computed. The value of the error bound characterizes the set of possible states of the preprocessed model, which is a norm ball with radius $\Delta(t)$ and center $Vx_{\mathrm{R}}(t)$. Since the states of the plant model are equal to the states of the preprocessed model multiplied by the inverse state transformation used in the preprocessing, the error bound also characterizes the set of possible states of the *plant* model, which is an *ellipsoid* in $\mathbb{R}^n$. The ellipsoid is projected onto each state of the plant to visualize the uncertainty of the quantities of interest in the figures below.

To study the error bound, the exemplary values $T_{\mathrm{in}} = 362.5\,\mathrm{K}$ and

$$u(t) := \begin{cases} 0\,\mathrm{K}/\mathrm{s} & \text{if } t < 8\,\mathrm{s}, \\ 0.06 \begin{bmatrix} 1 \\ -1 \\ 1 \end{bmatrix} T_{\mathrm{nom}}\,{}^{1}\!/\mathrm{s} & \text{if } t \geq 8\,\mathrm{s}. \end{cases} \tag{4.12}$$

are used. These values differ from the ones used to compute the snapshot to check the interpolation capabilities of the reduced model and error bound. The resulting error bounds for some important states of the plant model and the maximal fluid temperature inside the reactor are depicted in Figure 4.3. Since the three jacket temperatures and the maximal fluid temperature inside the reactor are constrained, the uncertainty of these values is important for the MPC scheme in the following chapters. For $\beta = 0.3$, the maximal uncertainty of the three jacket temperatures in Figure 4.3 is 4.6 K. For the maximal fluid temperature inside the reactor the uncertainty is below 2.5 K. Hence, the error bound is sufficiently small to allow the application within an MPC framework.

Figure 4.3 also depicts the error bound of Haasdonk and Ohlberger [2011], which relies on a bound of the norm of the matrix exponential with $\beta = 0$. As for $\beta = 0.3$, the LQR and a state transformation is used for the preprocessing. Since Problem 4.3 depends on β, the state transformation and also the shape of the ellipsoid depend on β. The error bound for $\beta = 0$ is tighter in general for a small time due to the smaller condition number of the state transformation matrix. The different shape of the ellipsoid explains why the error bound for $\beta = 0$ is nevertheless larger for the maximal fluid temperature for a small time.

Although the error bound with $\beta = 0.3$ is less tight for some variables on a short time interval, it outperforms the error bound of Haasdonk and Ohlberger [2011] for a longer simulation time. For MPC the state and input constraints have to be satisfied for the whole prediction horizon. Hence, the error bound should be tight for the time points when the constraints are active. A small error bound at the end of the prediction horizon can be crucial due to the terminal constraints. For the tubular reactor, the maximal temperature of the fluid inside the reactor and the terminal constraint are the critical constraints. Thus, for the prediction horizon of 16 s used in [Agudelo et al., 2007a], the generalization of the error bound presented in this thesis reduces the conservatism considerably.

For the comparison with [Dubljevic et al., 2006] a second reduced model with decoupled dominant states as in Remark 4.5 is used. For computing the reduced model and the error bound with the Schur decomposition the Multiprecision Computing Toolbox [Advanpix LLC, 2017] is used to increase the numerical accuracy. For the eigenvalue decomposition the preprocessed model is used since it leads to smaller error bounds compared to the original model. The dominant eigenvalues are chosen according to the Hankel singular values of the subsystems corresponding to each block of the dynamic matrix in real blockdiagonal form, see [Varga, 1995]. The Hankel singular values of the subsystems decay only by $3 \cdot 10^{-4}$ within the first 99 states. In comparison, the Hankel singular values of the whole preprocessed model decay by $1 \cdot 10^{-4}$ within the first 40 states. This indicates that for decoupling the dominant states a higher order of the reduced model is required for the example at hand.

In the following, we compare the reduced model of order 40 computed with the POD and the reduced model of order 99 with decoupled dominant states. The error in the maximal temperature between the plant and the reduced models as well as the error bounds are depicted in Figure 4.4 for $T_{\text{in}} = 362.5\,\text{K}$ and the input as in (4.12). For the reduced model with decoupled states the error bound

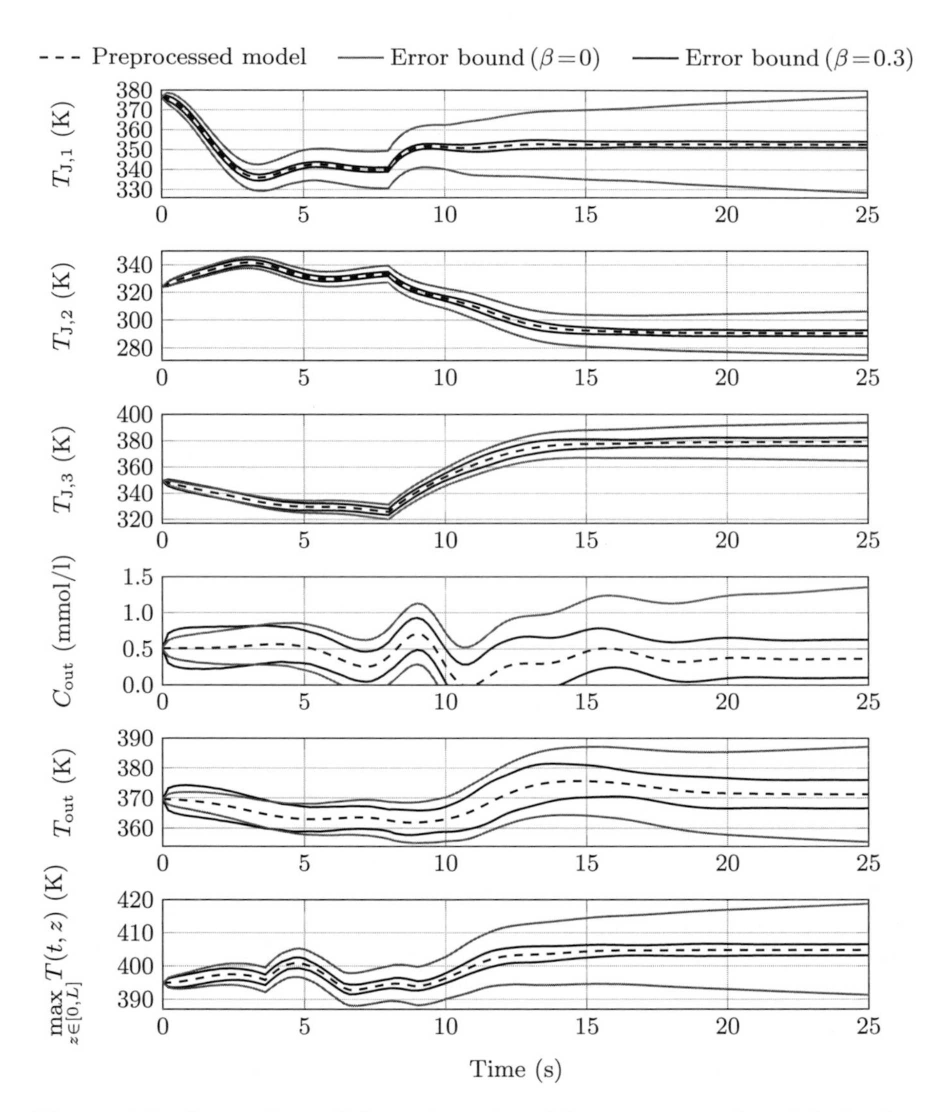

Figure 4.3: Comparison of the trajectories of the preprocessed model of order 303 with the lower and upper bound computed with the reduced model of order 40. The bounds are computed using the error bound presented in [Haasdonk and Ohlberger, 2011] ($\beta = 0$) and the error bound proposed in this work ($\beta = 0.3$).

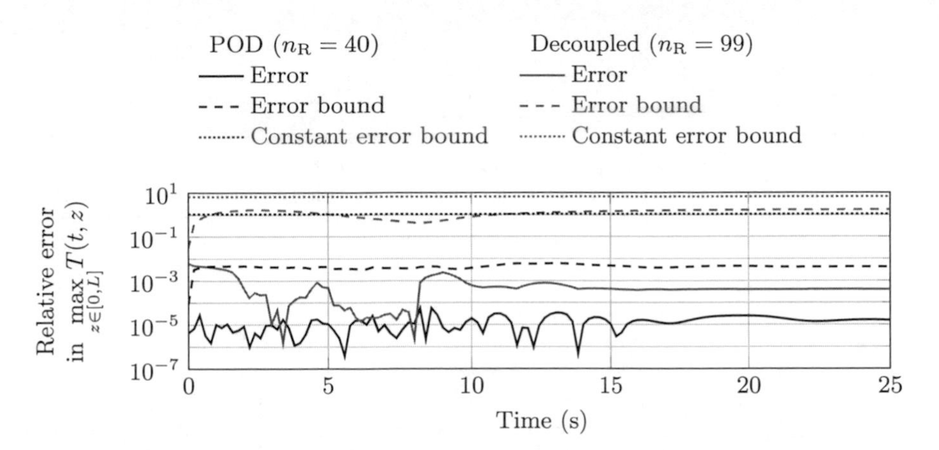

Figure 4.4: Model reduction error, the reduction error bound, and the constant reduction error bound for model reduction by POD and model reduction with decoupling the dominant states.

of Remark 4.5 has been used. Additionally, the constant error bounds for the worst-case states and inputs of Remark 4.6 are plotted. Since the states and inputs of the reduced models are not constrained to a compact set, we take the maximal value of the residual for the initial condition and inputs used to compute the snapshot in Section 4.5.2. To simplify the interpretation the values are scaled with the absolute norm of the maximal temperature of the plant.

Decoupling the dominant states results in a significant larger model reduction error than POD. This indicates that model reduction with decoupling the dominant states is inappropriate for the tubular reactor. The effect of bounding only the nondominant states becomes observable only for larger orders of the reduced model. Altogether, the error bound for the decoupled states is almost two orders of magnitude larger than POD even for more than twice the order of the reduced model. The constant error bound increases the error bound by at least two orders of magnitude for POD and around one order of magnitude for the decoupled dominant states. Hence, taking the actual input and states into account is significantly tighter. Comparing the error bound for the reduced model computed with POD with the constant error bound for the decoupled dominant states shows that the introduced error bound outperforms the error bound of [Dubljevic et al., 2006] by orders of magnitude for the tubular reactor.

4.5.4 Computational Demand

Statistical values of the computation time for simulation with forward sensitivity analysis of the preprocessed model, the reduced model, and the reduced model with

Table 4.1: Simulation time with forward sensitivity analysis.

Model	Average	Standard deviation
Preprocessed model (4.3)	75.30 s	0.23 s
Reduced model (4.6) without error bound	4.91 s	0.03 s
Reduced model (4.6) with error bound (4.7)	7.31 s	0.13 s

error bound for $\beta = 0.3$ are summarized in Table 4.1. For solving the ODEs we use the SUNDIALS toolbox [Hindmarsh et al., 2005]. The forward sensitivity analysis allows in the following chapter to compute the gradients of the cost functional and the constraints, which are used to solve the optimal control problem. One simulation with forward sensitivity analysis is on average 10 times faster for the reduced model with the error bounding system (4.7) than for the preprocessed model but less than 1.5 times slower than the reduced model without the error bound.

4.6 Summary

We presented an asymptotically stable error bounding system that provides guarantees for the set of possible states of the preprocessed model while simulating the reduced model. The error bounding system is derived in four steps. First, the plant is preprocessed by means of a prestabilizing feedback and a state transformation. Second, the norm of the matrix exponential of the preprocessed plant is bounded. Third, model reduction of the preprocessed model by projection is performed. Fourth, the bound for the norm of the matrix exponential is combined with the projection matrices defining the reduced model to obtain the error bounding system.

The preprocessing of the plant has several advantages. First, the preprocessing results in a modular approach separating the prestabilization and state transformation from the error bound and subsequent steps like the MPC approach discussed in the following chapters. Second, it introduces flexibility, which results in a tighter error bound. Finally, by the prestabilization an exponentially stable error bounding system is obtained even for unstable systems. For applications of the error bound where the input is fixed and no prestabilization is possible, an alternative to attain an asymptotically stable error bounding system is decoupling the unstable modes.

The proposed a-posteriori error bound generalizes the bound in [Haasdonk and Ohlberger, 2011] by adding an exponential decay rate to the bound for the norm of the matrix exponential. Due to the exponential decay rate, the improved error bound is considerably tighter than the original one especially for a long simulation time. At the same time the error bound possesses a comparable computational demand. The proposed error bound takes the actual state and inputs into account and is significantly tighter for the considered tubular reactor than an error bound for the worst-case states and inputs.

Chapter 5

Model Predictive Control Using Reduced Models with Guaranteed Properties for Continuous-Time Systems

In the previous chapter, an a-posteriori bound for the model reduction error has been introduced. The goal of this chapter is to derive an MPC scheme using the reduced model that utilizes the error bound to give rigorous guarantees for the closed loop with the high-dimensional plant. As seen in Section 1.2.3 only a few and limited results exist in this research field even for linear systems. To overcome these limitations, we also concentrate on the common problem setup of LTI systems, polytopic constraints, and a quadratic stage cost. For this problem setup we show the advantage of the derived method compared to the already existing methods in the context of MPC using reduced models.

The problem setup is stated in Section 5.1 together with an academic example motivating the need of guarantees for the closed loop with the plant model when using a reduced model for the prediction. In Section 5.2, the preprocessing of the plant model and the model reduction is applied to the problem setup at hand such that the a-posteriori error bound of the previous chapter is applicable. The error bound is used in Section 5.3 to ensure constraint satisfaction for the preprocessed model in a computationally efficient way. Then, in Section 5.4, the MPC scheme including the error bounding system is presented. In Section 5.5, we show that for a certain choice of design parameters the model reduction error can be eliminated in the cost functional. Subsequently, we discuss the implications for the proposed MPC scheme. Furthermore, we introduce a generalized assumption for the stage cost to allow also for the introduced cost functional. For this generalized assumption, we prove recursive feasibility and asymptotic stability of the preprocessed model in closed loop with the proposed MPC scheme in Section 5.6. In Section 5.7, we compare the derived MPC scheme with approaches existing in the literature. Finally, in Section 5.8, we apply the proposed MPC scheme to the tubular reactor and compare the performance with the two MPC schemes introduced in Section 2.3.

Most parts of this chapter are based on [Löhning et al., 2014]. The main differences to [Löhning et al., 2014] are the preprocessing of the plant instead of the feedback of the model reduction error, Lemma 5.11 about the elimination of the model reduction error in the cost functional, and the consequences of both changes for the theorems and the application to the tubular reactor.

5.1 Problem Statement

In this chapter, we consider continuous-time LTI plants described by

$$\Sigma_{\mathrm{P}}: \begin{cases} \dot{x}_{\mathrm{P}}(t) = A_{\mathrm{P}} x_{\mathrm{P}}(t) + B_{\mathrm{P}} u_{\mathrm{P}}(t)\,, \\ x_{\mathrm{P}}(0) = x_{\mathrm{P},0}\,, \end{cases} \tag{5.1}$$

in which $x_{\mathrm{P}}(t) \in \mathbb{R}^n$ is the state vector at time t, $u_{\mathrm{P}}(t) \in \mathbb{R}^{n_{\mathrm{U}}}$ is the input vector at time t, $u_{\mathrm{P}}(\cdot) \in \mathcal{L}_2^{n_{\mathrm{U}}}$, and $x_{\mathrm{P},0} \in \mathbb{R}^n$ is the initial condition. The model order n is assumed to be large, i.e., $n \gg 1$. The plant satisfies the following assumption.

Assumption 5.1. *The pair $(A_{\mathrm{P}}, B_{\mathrm{P}})$ is stabilizable.*

Furthermore, we assume that the full state of the plant can be measured. This assumption can be relaxed as shown later in Proposition 5.10.

The states and inputs are constrained to a polytopic set by

$$C_{\mathrm{P}} \begin{bmatrix} x_{\mathrm{P}}(t) \\ u_{\mathrm{P}}(t) \end{bmatrix} \leq d_{\mathrm{P}}\,, \tag{5.2}$$

in which $C_{\mathrm{P}} \in \mathbb{R}^{n_{\mathrm{C}} \times n+n_{\mathrm{U}}}$ and $d_{\mathrm{P}} \in \mathbb{R}^{n_{\mathrm{C}}}$ for all $t \geq 0$. To prove asymptotic stability in Section 5.6, we state the following common assumption.

Assumption 5.2. *The state and input constraints* (5.2) *define a compact set and the equilibrium $x_{\mathrm{P}} = 0$ for $u_{\mathrm{P}} = 0$ is contained in its interior.*

Given the initial state $x_{\mathrm{P},0}$, we want to steer the system to the origin close to optimality with respect to the infinite horizon cost functional

$$J_{\mathrm{P}}^{\mathrm{inf}}(x_{\mathrm{P},0}, u_{\mathrm{P}}) := \int_0^\infty F_{\mathrm{P}}\big(x_{\mathrm{P}}(t), u_{\mathrm{P}}(t)\big)\,\mathrm{d}t\,, \tag{5.3}$$

subject to the system dynamics (5.1) as well as the state and input constraints (5.2). A common cost function is the quadratic stage cost, which is defined by

$$F_{\mathrm{P}}\big(x_{\mathrm{P}}(t), u_{\mathrm{P}}(t)\big) := x_{\mathrm{P}}^{\mathsf{T}}(t) Q_{\mathrm{P}} x_{\mathrm{P}}(t) + u_{\mathrm{P}}^{\mathsf{T}}(t) R_{\mathrm{P}} u_{\mathrm{P}}(t)\,,$$

in which Q_{P} and R_{P} satisfy the following assumption.

Assumption 5.3. *The matrices Q_{P} and R_{P} are symmetric and positive definite.*

Assumption 5.3 is used in Lemma 5.11 and in the following sections as a sufficient condition for the relaxed Assumption 5.16 about the cost function parameters for the reduced model. The proof of asymptotic stability in Section 5.6 relies on Assumption 5.16. Hence, Assumption 5.3 is only sufficient but not necessary.

The given control problem without the constraints (5.2) can be solved using the well-known LQR [Kwakernaak and Sivan, 1972]. Due to the constraints it is significantly harder to obtain a good performance of the controlled system. Thus, we seek an approximate solution via MPC. To solve the optimal control problem, the plant model is frequently replaced by a reduced model as discussed in Section 2.3.4.

The model reduction error can have a significant influence on the closed-loop behavior of the common approach for MPC using a reduced model presented in Section 2.3.4, as we will see in the following example.

Example 5.4. *Consider the plant*

$$\dot{x}_\mathrm{P}(t) = \begin{bmatrix} 0.1 & 0.7125 \\ 0 & -0.9 \end{bmatrix} x_\mathrm{P}(t) + \begin{bmatrix} 0.2 \\ -0.19 \end{bmatrix} u_\mathrm{P}(t), \qquad x_\mathrm{P}(0) = x_{\mathrm{P},0},$$

with the constraints $|x_{\mathrm{P},1}(t)| \leq 1$, $|x_{\mathrm{P},2}(t)| \leq 10$, and $|u_\mathrm{P}(t)| \leq 0.5$. The stage cost is given by $Q_\mathrm{P} = I_2$ and $R_\mathrm{P} = 1$.

Modal reduction by modal truncation would be a good choice for this plant model containing one unstable and another nondominant mode. To show the influence of the model reduction error for model reduction by projection in general, i.e., with a coupling between the states of the reduced model and the neglected states, consider $V = W = [1 \quad 0]^\mathsf{T}$, which results in a reduced model that contains only the unstable state.

For $\mathcal{P}$-MPC the terminal set is the maximal positively invariant polytope for the closed loop of the plant and the LQR. The terminal cost is $E_\mathcal{P}(x) = x^\mathsf{T} Q_\mathcal{P}^\Omega x$ in which $Q_\mathcal{P}^\Omega$ is the solution of the algebraic Riccati equation (ARE) associated with the plant model. For $\mathcal{R}$-MPC the terminal set and terminal cost is computed analogous to $\mathcal{P}$-MPC but using the reduced model.

Trajectories of the closed loop for a sampling time $\delta = 0.2$, a prediction horizon $T = 50\,\delta$, and a piecewise constant input are shown in Figure 5.1. While the plant in closed loop with $\mathcal{P}$-MPC converges to the origin, the closed loop with $\mathcal{R}$-MPC diverges due to the model reduction error.

The example underpins the lack of guaranteed stability when the plant model is replaced with a reduced model for the prediction. To recover the guarantees of $\mathcal{P}$-MPC stated in Section 2.3.3, i.e., constraint satisfaction, recursive feasibility, and asymptotic stability, the $\mathcal{R}$-MPC scheme is extended in this chapter with the a-posteriori error bound derived in Chapter 4. Motivated by the symbol for the error bound Δ, the resulting MPC scheme using the reduced model and the error bounding system is denoted Δ-MPC.

The MPC schemes considered in this thesis are visualized in Figure 5.2 and identified in Table 5.1. For details of the Δ-MPC scheme we refer to the subsequent sections.

5.2 Preprocessing and Model Reduction

To apply the a-posteriori error bound of Theorem 4.4, the plant model is preprocessed as presented in Section 4.2. This results in the preprocessed model. Thereby, an asymptotically stable error bounding system can be achieved also for

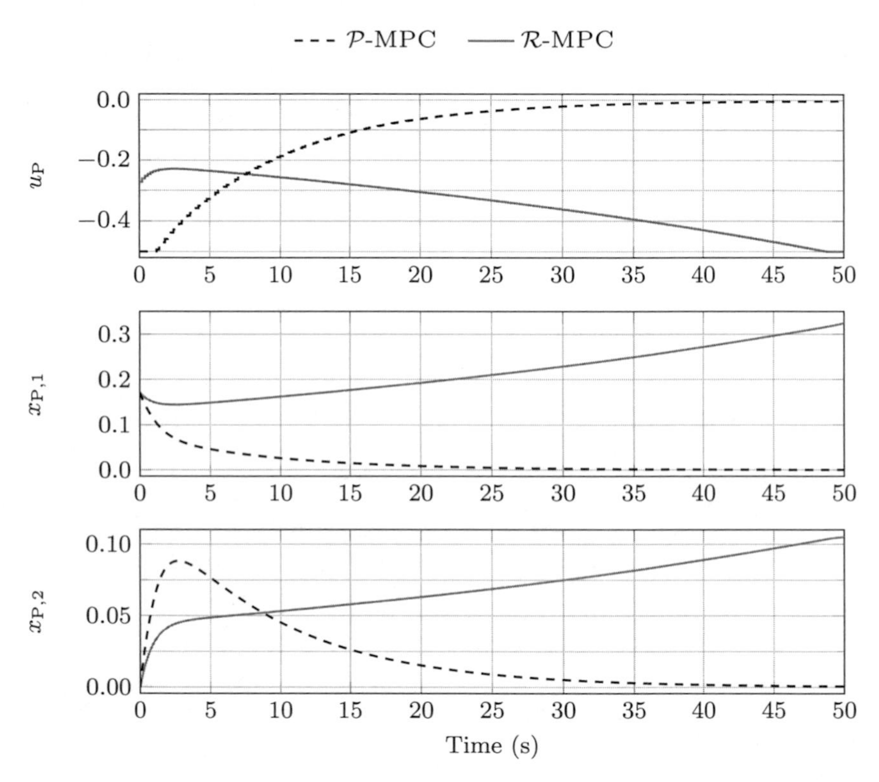

Figure 5.1: Trajectories of the plant in closed loop with MPC using the plant model ($\mathcal{P}$-MPC) as well as MPC using the reduced model ($\mathcal{R}$-MPC).

Table 5.1: MPC schemes considered in this thesis.

MPC scheme	Prediction model	Input of the plant	Optimization problem
$\mathcal{P}$-MPC	Σ_{P}, see (2.6)	$u_{\mathcal{P}}(t)$, see (2.10)	Problem 2.7
$\mathcal{R}$-MPC	Σ_{R} derived from Σ_{P}, see (2.12a)–(2.12b)	$u_{\mathcal{R}}(t)$, see (2.13)	Problem 2.13
Δ-MPC	Σ_{R} and Σ_{Δ}, see (5.7) and (5.8)	$u_{\Delta}(t) - K_{\mathrm{P}} x_{\mathrm{P}}(t)$, see (5.13)	Problem 5.8

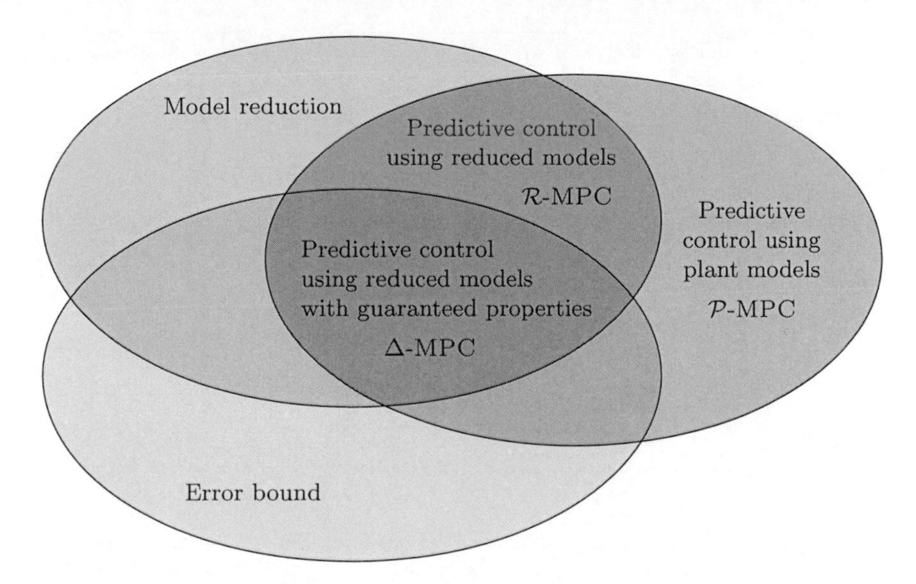

Figure 5.2: Visualization of the MPC schemes considered in this thesis.

unstable plant models. Afterwards, model reduction and MPC is applied to the preprocessed model. The structure of the proposed MPC scheme is illustrated in Figure 5.3. The overall controller consists of the model predictive controller and the preprocessing of the plant model. The model predictive controller includes the reduced model as well as the scalar error bounding system.

In the remainder of this section, we apply the preprocessing presented in Section 4.2 and model reduction by projection to the problem setup.

The prestabilization of the plant $u_{\mathrm{P}}(t) = u(t) - K_{\mathrm{P}} x_{\mathrm{P}}(t)$ and the linear state transformation $x(t) = T_{\mathrm{P}} x_{\mathrm{P}}(t)$ with $T_{\mathrm{P}} \succ 0$ results in the preprocessed model

$$\Sigma_{\mathrm{PP}} : \begin{cases} \dot{x}(t) = Ax(t) + Bu(t) \,, \\ x(0) = x_0 \,, \end{cases} \tag{5.4}$$

in which $A = T_{\mathrm{P}}(A_{\mathrm{P}} - B_{\mathrm{P}} K_{\mathrm{P}}) T_{\mathrm{P}}^{-1}$, $B = T_{\mathrm{P}} B_{\mathrm{P}}$, and $x_0 = T_{\mathrm{P}} x_{\mathrm{P},0}$. For every stabilizable plant there exists a feedback K_{P} such that the preprocessed model is asymptotically stable. The state and input constraints for the preprocessed model are

$$C \begin{bmatrix} x(t) \\ u(t) \end{bmatrix} \leq d \,, \tag{5.5}$$

in which

$$C = C_{\mathrm{P}} \begin{bmatrix} T_{\mathrm{P}}^{-1} & 0_{n \times n_{\mathrm{U}}} \\ -K_{\mathrm{P}} T_{\mathrm{P}}^{-1} & I_{n_{\mathrm{U}}} \end{bmatrix}$$

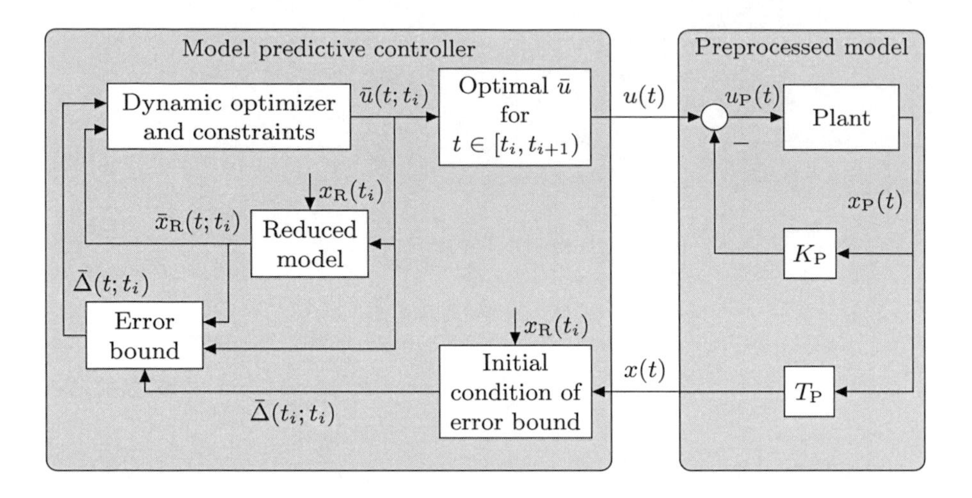

Figure 5.3: Structure of the proposed MPC scheme. The update of the sampling instant t_i is omitted. For details we refer to Algorithm 5.9. The initial condition of the reduced model $x_{\mathrm{R}}(t_i)$ is discussed in Section 5.4.

and $d = d_{\mathrm{P}}$. The infinite horizon cost functional for the preprocessed model is given by

$$J^{\mathrm{inf}}(x_0, u) := \int_0^\infty F\big(x(t), u(t)\big)\,\mathrm{d}t\,,$$

with the quadratic stage cost

$$F\big(x(t), u(t)\big) := x^{\mathsf{T}}(t)Qx(t) + 2x^{\mathsf{T}}(t)Su(t) + u^{\mathsf{T}}(t)Ru(t)\,, \tag{5.6}$$

in which $Q = T_{\mathrm{P}}^{-\mathsf{T}} Q_{\mathrm{P}} T_{\mathrm{P}}^{-1} + T_{\mathrm{P}}^{-\mathsf{T}} K_{\mathrm{P}}^{\mathsf{T}} R_{\mathrm{P}} K_{\mathrm{P}} T_{\mathrm{P}}^{-1}$, $S = -T_{\mathrm{P}}^{-\mathsf{T}} K_{\mathrm{P}}^{\mathsf{T}} R_{\mathrm{P}}$, and $R = R_{\mathrm{P}}$.

Assumptions 5.2 and 5.3 are equivalent to the following assumptions.

Assumption 5.5. *The state and input constraints* (5.5) *define a compact set and the equilibrium $x = 0$ for $u = 0$ is contained in its interior.*

Assumption 5.6. *The matrices $Q - SR^{-1}S^{\mathsf{T}}$ and R are symmetric and positive definite.*

As for the error bound derived in Section 4.4, we allow for any model reduction method based on a projection of the preprocessed model as introduced in Section 2.1.1. Therefore, we assume that the matrices V and W are given and result in the reduced model

$$\Sigma_{\mathrm{R}} : \begin{cases} \dot{x}_{\mathrm{R}}(t) = A_{\mathrm{R}} x_{\mathrm{R}}(t) + B_{\mathrm{R}} u(t)\,, \\ x_{\mathrm{R}}(0) = x_{\mathrm{R},0}\,, \end{cases} \tag{5.7}$$

in which $A_{\mathrm{R}} = W^{\mathsf{T}}AV$ and $B_{\mathrm{R}} = W^{\mathsf{T}}B$. When using the reduced model for the prediction, the estimated state $Vx_{\mathrm{R}}(t)$ is used in the stage cost

$$F_{\mathrm{R}}(x_{\mathrm{R}}(t), u(t)) := F(Vx_{\mathrm{R}}(t), u(t)) = x_{\mathrm{R}}^{\mathsf{T}}(t)Q_{\mathrm{R}}x_{\mathrm{R}}(t) + 2x_{\mathrm{R}}^{\mathsf{T}}(t)S_{\mathrm{R}}u(t) + u^{\mathsf{T}}(t)Ru(t)\,,$$

in which $Q_{\mathrm{R}} = V^{\mathsf{T}}QV$, $S_{\mathrm{R}} = V^{\mathsf{T}}S$.

5.3 Guaranteeing Constraint Satisfaction

To enforce the state and input constraints (5.5), we extend the $\mathcal{R}$-MPC scheme given by Algorithm 2.14 with the error bounding system

$$\Sigma_\Delta : \begin{cases} \dot{\bar{\Delta}}(t;t_i) = -\beta\bar{\Delta}(t;t_i) + \alpha\left\| \left(I_n - VW^{\mathsf{T}}\right)\left(AV\bar{x}_{\mathrm{R}}(t;t_i) + B\bar{u}(t;t_i)\right)\right\|, \\ \bar{\Delta}(t_i;t_i) = \alpha\|x(t_i) - Vx_{\mathrm{R}}(t_i)\| \end{cases} \tag{5.8}$$

stated in Theorem 4.4. Based on the error bounding system, we know the set of possible states of the preprocessed model, i.e.,

$$x(t) \in \mathcal{X}\left(\bar{x}_{\mathrm{R}}(t;t_i), \bar{\Delta}(t;t_i)\right) = \left\{x \in \mathbb{R}^n \mid \|x - V\bar{x}_{\mathrm{R}}(t;t_i)\| \leq \bar{\Delta}(t;t_i)\right\}.$$

The bar over x_{R} and Δ denotes that the signal is predicted within the model predictive controller. The set of possible states of the preprocessed model allows to enforce the state and input constraints for the preprocessed model (5.5) by using the constraints

$$C\begin{bmatrix} x(t) \\ \bar{u}(t;t_i) \end{bmatrix} \leq d \quad \text{for all } x(t) \in \mathcal{X}\left(\bar{x}_{\mathrm{R}}(t;t_i), \bar{\Delta}(t;t_i)\right) \tag{5.9}$$

in the optimization problem. To check if a set is contained in a polytope is computationally expensive for general sets. Fortunately, the set $\mathcal{X}$ is an Euclidean ball with the center $V\bar{x}_{\mathrm{R}}(t;t_i)$ and radius $\bar{\Delta}(t;t_i)$. Hence, the constraints (5.9) can be formulated as polytopic constraints in $\bar{x}_{\mathrm{R}}(t;t_i)$, $\bar{u}(t;t_i)$, and $\bar{\Delta}(t;t_i)$, i.e., without the set $\mathcal{X}$, as shown by the following proposition.

Proposition 5.7 (Reformulation of constraints). *The constraints* (5.9) *are equivalent to*

$$c_k^{\mathsf{T}}\begin{bmatrix} V\bar{x}_{\mathrm{R}}(t;t_i) \\ \bar{u}(t;t_i) \end{bmatrix} \leq d_k - \bar{\Delta}(t;t_i)\left\| c_k^{\mathsf{T}}\begin{bmatrix} I_n \\ 0_{n_{\mathrm{U}}\times n} \end{bmatrix}\right\|, \qquad k = 1, \ldots, n_{\mathrm{C}}. \tag{5.10}$$

in which c_k^{T} is the k-th row of C and d_k the k-th element of d.

Proof. The constraints (5.9) are equivalent to: For every $k = 1, \ldots, n_{\mathrm{C}}$

$$c_k^{\mathsf{T}}\begin{bmatrix} V\bar{x}_{\mathrm{R}}(t;t_i) \\ \bar{u}(t;t_i) \end{bmatrix} + c_k^{\mathsf{T}}\begin{bmatrix} I_n \\ 0_{n_{\mathrm{U}}\times n} \end{bmatrix} e_k(t) \leq d_k \quad \text{for all } e_k(t) \in \mathbb{R}^n \text{ with } \|e_k(t)\| \leq \bar{\Delta}(t;t_i)\,. \tag{5.11}$$

For every k with $c_k^\mathsf{T}\begin{bmatrix} I_n & 0_{n\times n_\mathrm{U}}\end{bmatrix}^\mathsf{T} = 0_{n_\mathrm{C}\times n}$, equivalence of (5.10) and (5.11) follows directly. For all other k, from the upper bound of the norm of $e_k(t)$ follows that the left side of the inequality in (5.11) is maximal if and only if $e_k(t)$ is parallel to $\begin{bmatrix} I_n & 0_{n\times n_\mathrm{U}}\end{bmatrix} c_k$ and points in the same direction. This results in

$$e_k(t) = \frac{\bar{\Delta}(t;t_i)}{\left\|\begin{bmatrix} I_n & 0_{n\times n_\mathrm{U}}\end{bmatrix} c_k\right\|}\begin{bmatrix} I_n & 0_{n\times n_\mathrm{U}}\end{bmatrix} c_k\,.$$

Consequently, the constraints (5.10) and (5.11) are equivalent. Altogether, the constraints (5.9) and (5.10) are equivalent. □

Up to now, we are able to satisfy the state and input constraints (5.5) while using the reduced model for the prediction.

5.4 MPC Scheme Using the Reduced Model and Error Bound

In this section, we derive the optimization problem of the MPC scheme using the reduced model and the error bounding system such that asymptotic stability of the closed loop can be guaranteed later in Section 5.6 by an appropriate choice of the terminal cost and terminal set.

For the application of the model predictive controller and to prove asymptotic stability of the closed loop, it is important that the underlying optimal control problem is feasible for all time instants. This would be guaranteed if, first, a feasible solution at $t_0 = 0$ exists and if, second, from feasibility at $t_0 = 0$ follows feasibility for all subsequent sampling instants when applying the model predictive controller.

In general, feasibility at the sampling instant $t_{i+1} = t_i + \delta$ is proven by defining a candidate solution based on the solution at the previous sampling instant t_i. By assumption, the solution of the sampling instant t_i is feasible for $t \in [t_{i+1}, t_i + T]$ and this solution is extended until $t_{i+1} + T$ based on a local control law, which is designed such that certain properties are satisfied in the terminal set as in Assumption 2.11.

In the optimization problem for $\mathcal{R}$-MPC, i.e., Problem 2.13, the initial condition of the reduced model is the projection of the measured high-dimensional state. For this choice, recursive feasibility is not guaranteed as illustrated by example in Figure 5.4. The prediction for the state of the reduced model at t_{i+1} based on time t_i, i.e., $\bar{x}_\mathrm{R}(t_{i+1};t_i)$ and the predicted error bound $\bar{\Delta}(t_{i+1};t_i)$ define the set of possible states of the preprocessed model. The actual state of the preprocessed model $x(t_{i+1})$ is in this set. The projected state of the preprocessed model $W^\mathsf{T}x(t_{i+1})$ and the predicted state of the reduced model $\bar{x}_\mathrm{R}(t_{i+1};t_i)$ are not equal. When using the projected state as initial condition for the reduced model, the set of possible states at the sampling instant t_{i+1}, i.e., $\mathcal{X}\big(W^\mathsf{T}x(t_{i+1}), \left\|(I_n - VW^\mathsf{T})x(t_{i+1})\right\|\big)$ is not contained in the set predicted at the sampling instant t_i, i.e., $\mathcal{X}\big(\bar{x}_\mathrm{R}(t_{i+1};t_i), \bar{\Delta}(t_{i+1};t_i)\big)$.

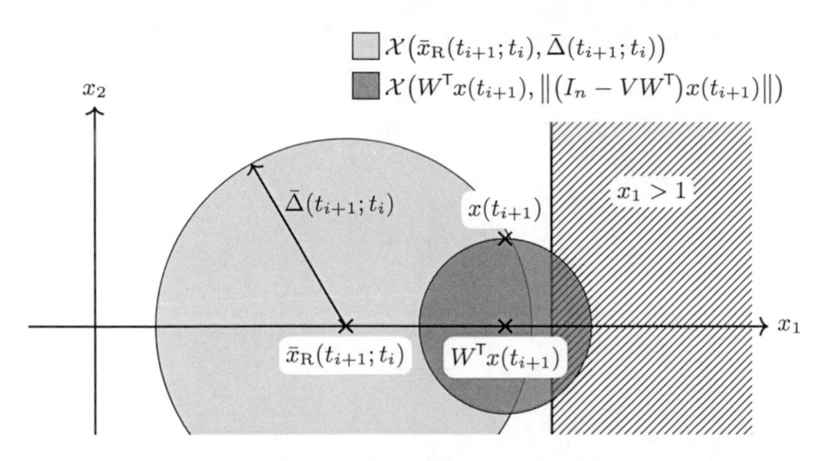

Figure 5.4: Illustration of the set predicted at the sampling instant t_i(□) and the set for the possible state of the preprocessed model at the sampling instant t_{i+1} (■) for $n = 2$, $V = W = \begin{bmatrix}1 & 0\end{bmatrix}^{\mathsf{T}}$, and $\alpha = 1$ when the reduced model is initialized with the projected state of the preprocessed model, i.e., $x_{\mathrm{R}}(t_{i+1}) = W^{\mathsf{T}}x(t_{i+1})$.

This is possible even for $\alpha = 1$ where the smallest possible initial condition of the error bound is attained. The optimization problem for the example depicted in Figure 5.4 is not feasible at the sampling instant t_{i+1} as visualized by the state constraint $x_1 \leq 1$.

Thus, using the projected state of the preprocessed model as initial condition of the reduced model might jeopardize recursive feasibility of the MPC scheme. One possibility to get recursive feasibility is to use the initial state of the reduced model as additional decision variable, which is also used in the area of robust MPC in order to deal with a model plant mismatch [Mayne et al., 2005]. To have fewer decision variables, we restrict $x_{\mathrm{R}}(t_{i+1})$ to the set $\mathcal{I} := \left\{x \in \mathbb{R}^{n_{\mathrm{R}}} \mid x = \gamma_1 W^{\mathsf{T}}x(t_{i+1}) + \gamma_2 \bar{x}_{\mathrm{R}}(t_{i+1};t_i)\right\}$ with $\gamma_1, \gamma_2 \in \mathbb{R}$, i.e., the subspace containing the origin, the projection of the state of the preprocessed model, and the predicted state of the reduced model. Allowing for $x_{\mathrm{R}}(t_{i+1}) = \bar{x}_{\mathrm{R}}(t_{i+1};t_i)$ is exploited to prove recursive feasibility later on. The projection of the state of the preprocessed model is included in $\mathcal{I}$ to allow for an adaption of $x_{\mathrm{R}}(t_{i+1})$ towards the measured state of the preprocessed model. Why the subspace also contains the origin will become clear later in Section 5.5. Furthermore, this simplifies the proof of asymptotic stability.

For the proposed MPC approach using the reduced model and error bounding system, at every sampling instant $t_i = i\delta$, $i \in \mathbb{N}_0$ the following finite horizon optimal control problem is solved.

Problem 5.8 (Optimization problem for MPC using the reduced model and error bounding system).

$$\underset{\substack{\bar{u}\in\mathcal{PC}^{n_\mathrm{U}}_{[t_i,t_i+T]}\\ x_\mathrm{R}(t_i)\in\mathcal{I},\bar{x}_\mathrm{R},\bar{\Delta}}}{\text{minimize}} \quad J_\Delta\big(x_\mathrm{R}(t_i),\bar{u}(\cdot;t_i)\big) := \int_{t_i}^{t_i+T} F_\mathrm{R}\big(\bar{x}_\mathrm{R}(t;t_i),\bar{u}(t;t_i)\big)\,\mathrm{d}t + E_\Delta\big(\bar{x}_\mathrm{R}(t_i+T;t_i)\big)$$

subject to

$$\dot{\bar{x}}_\mathrm{R}(t;t_i) = A_\mathrm{R}\bar{x}_\mathrm{R}(t;t_i) + B_\mathrm{R}\bar{u}(t;t_i)\,, \tag{5.12a}$$

$$\bar{x}_\mathrm{R}(t_i;t_i) = x_\mathrm{R}(t_i)\,, \tag{5.12b}$$

$$c_k^\mathsf{T}\begin{bmatrix} V\bar{x}_\mathrm{R}(t;t_i)\\ \bar{u}(t;t_i)\end{bmatrix} \le d_k - \bar{\Delta}(t;t_i)\left\| c_k^\mathsf{T}\begin{bmatrix} I_n\\ 0_{n_\mathrm{U}\times n}\end{bmatrix}\right\|, \qquad k=1,\ldots,n_\mathrm{C}\,, \tag{5.12c}$$

$$\big(\bar{x}_\mathrm{R}(t_i+T;t_i),\bar{\Delta}(t_i+T;t_i)\big) \in \Omega_\Delta\,, \tag{5.12d}$$

$$\dot{\bar{\Delta}}(t;t_i) = -\beta\bar{\Delta}(t;t_i) + \alpha\Big\|\Big(I_n - VW^\mathsf{T}\Big)\big(AV\bar{x}_\mathrm{R}(t;t_i)+B\bar{u}(t;t_i)\big)\Big\|\,, \tag{5.12e}$$

$$\bar{\Delta}(t_i;t_i) = \begin{cases} \Delta^*(t_i;t_{i-1}) & \text{if } x_\mathrm{R}(t_i) = x^*_\Delta(t_i;t_{i-1}) \text{ and}\\ & \Delta^*(t_i;t_{i-1}) < \alpha\|x(t_i) - Vx_\mathrm{R}(t_i)\|\,,\\ \alpha\|x(t_i)-Vx_\mathrm{R}(t_i)\| & \text{otherwise}\,, \end{cases} \tag{5.12f}$$

for all $t\in[t_i,t_i+T]$.

Problem 5.8 contains, in addition to the reduced model (5.12a)–(5.12b), the error bounding system (5.12e)–(5.12f). In comparison to the optimization problem of $\mathcal{R}$-MPC, the constraints (5.12c) are tightened by the error bound. Furthermore, the error bound is included in the terminal constraint (5.12d), which allows to guarantee satisfaction of the constraints within the terminal set. The input trajectory and initial condition of the reduced model that solve Problem 5.8 are denoted by $u^*_\Delta(t;t_i)$ and $x^*_\mathrm{R}(t_i)$, with associated predicted state of the reduced model $x^*_\Delta(t;t_i)$ and predicted error bound $\Delta^*(t;t_i)$ for $t\in[t_i,t_i+T]$. The cost associated with the optimal solution is denoted by $J^*_\Delta\big(x^*_\mathrm{R}(t_i),u^*_\Delta(\cdot;t_i)\big)$. The subscript Δ denotes that the error bounding system is used (together with the reduced model) for the prediction.

For $\alpha>1$, at each sampling instant the initial error bound in (5.8) is overestimated by the factor α. To reduce the conservatism, the error bounding system is initialized in (5.12f) with the error bound predicted at the last sampling instant if the reduced model is initialized with the state predicted at the last sampling instant and the error bound is decreased. Furthermore, we use this initialization of the error bound to establish recursive feasibility in Section 5.6. When solving Problem 5.8, the distinction of cases in (5.12f) can be circumvented by solving two optimization problems, one for each case in (5.12f), and using the solution with the lower cost value. If $\alpha=1$, then the error bounding system guarantees

$\alpha\|x(t_i) - Vx^*_\Delta(t_i;t_{i-1})\| \leq \Delta^*(t_i;t_{i-1})$. In this case, the first case in (5.12f) cannot occur and the distinction of cases disappears in (5.12f). Due to (5.12f) and the set of initial states $\mathcal{I}$, Problem 5.8 does not only depend on $x(t_i)$, but also on $x^*_\Delta(t_i;t_{i-1})$ and $\Delta^*(t_i;t_{i-1})$. For $t_i = 0$, we use $\Delta^*(0;0) := \alpha\|x(0) - Vx_{\mathrm{R}}(0)\|$ and $x_{\mathrm{R}}(0) \in \{x \in \mathbb{R}^{n_{\mathrm{R}}} \mid x = \gamma_1 W^{\mathsf{T}} x_0\}$ with $\gamma_1 \in \mathbb{R}$.

The optimal input is applied to the preprocessed model until the next sampling instant $t_{i+1} = t_i + \delta$. Hence,

$$u(t) = u_\Delta(t) := u^*_\Delta(t;t_i)\,, \qquad t_i \leq t < t_{i+1}\,. \tag{5.13}$$

The MPC scheme using the reduced model and error bounding system is denoted Δ-MPC and is given by the following algorithm.

Algorithm 5.9 (Δ-MPC: MPC using the reduced model and error bounding system).

Require: *The preprocessed model* Σ_{PP} (5.4), *the matrices* V *and* W *defining the reduced model* Σ_{R} (5.7), *the constraints* (5.5), *the error bounding system* Σ_Δ (5.8) *with* $\alpha \geq 1$ *and* $\beta < 0$, *the sampling time* δ *with* $0 < \delta \leq T$, *the matrices defining the stage cost* Q_{R}, S_{R}, R, *the terminal cost* $E_\Delta(\cdot)$, *and the terminal set* Ω_Δ

$i \leftarrow 0$

loop

$x(t_i) \leftarrow$ *Measure the state of the preprocessed model at time* t_i

$u^*_\Delta(\cdot;t_i)$, $x^*_\Delta(t_{i+1};t_i)$, $\Delta^*(t_{i+1};t_i) \leftarrow$ *Solve Problem 5.8*

$u(\cdot) = u_\Delta(\cdot) \leftarrow$ *Apply* $u^*_\Delta(t;t_i)$ *for* $t \in [t_i, t_{i+1})$ *to* Σ_{PP}

$i \leftarrow i+1$

end loop

In reality, the state of the plant and not the state of the preprocessed model is measured. However, to ease the presentation, Algorithm 5.9 and the subsequent theorems are formulated for the preprocessed model.

Measuring the full state of the preprocessed model or using an observer to reconstruct the full state of the preprocessed model is difficult for a high-dimensional system in general. When an appropriate prestabilization is used, the assumption of the full state measurement can be relaxed as stated in the following proposition.

Proposition 5.10 (Partial state measurement). *Consider the* SVD *of the full-column rank matrix* $V = \check{U}\check{\Sigma}\check{V}^{\mathsf{T}}$, *in which* $\check{U}^{\mathsf{T}}\check{U} = I_{n_{\mathrm{R}}}$, $\check{\Sigma} = \mathrm{diag}\left(\sigma_1(V), \ldots, \sigma_{n_{\mathrm{R}}}(V)\right)$, *and* $\check{V}^{\mathsf{T}}\check{V} = I_{n_{\mathrm{R}}}$. *Furthermore, consider a prestabilization of the form*

$$u_{\mathrm{P}}(t) = u(t) - K\check{U}^{\mathsf{T}} T_{\mathrm{P}} x_{\mathrm{P}}(t)$$

is used and $A_{\mathrm{P}} - B_{\mathrm{P}} K \check{U}^{\mathsf{T}} T_{\mathrm{P}}$ *is Hurwitz. Then, a measurement of the orthogonal projection of* $x(t) = T_{\mathrm{P}} x_{\mathrm{P}}(t)$ *onto the image of* V *together with an upper bound for the norm of the orthogonal projection of* $x(t) = T_{\mathrm{P}} x_{\mathrm{P}}(t)$ *onto the kernel of* V^{T} *is sufficient for* Δ-MPC.

Proof. For Δ-MPC, the state of the preprocessed model is only required for the prestabilization and in Problem 5.8 to initialize the error bounding system. The prestabilization is assumed to depend only on the orthogonal projection of $x(t) = T_\mathrm{P} x_\mathrm{P}(t)$ onto the image of V. For the initialization of the error bound consider

$$\begin{aligned}\|x(t_i) - V x_\mathrm{R}(t_i)\|^2 &= x^\mathsf{T}(t_i)x(t_i) - 2x^\mathsf{T}(t_i)\check{U}\check{\Sigma}\check{V}^\mathsf{T} x_\mathrm{R}(t_i) + x_\mathrm{R}^\mathsf{T}(t_i)\check{V}\check{\Sigma}^2\check{V}^\mathsf{T} x_\mathrm{R}(t_i)\\ &= \left\|\check{U}^\mathsf{T} x(t_i) - \check{\Sigma}\check{V}^\mathsf{T} x_\mathrm{R}(t_i)\right\|^2 + \left\|\left(I_n - \check{U}\check{U}^\mathsf{T}\right)x(t_i)\right\|^2.\end{aligned} \tag{5.14}$$

Hence, the initialization of the error bounding system can be divided into a first term depending only on the orthogonal projection of $x(t) = T_\mathrm{P} x_\mathrm{P}(t)$ onto the image of V and a second term that is the norm of the orthogonal projection of $x(t) = T_\mathrm{P} x_\mathrm{P}(t)$ onto the kernel of V^T. To conclude the proof, note that the constraint satisfaction and recursive feasibility remain true if an upper bound is used for the initial value of the error bound. □

The full-rank factorization [Piziak and Odell, 1999] can be computed with the SVD.

The reformulation of the initialization of the error bounding system presented in (5.14) results in a low-dimensional inequality constraint because the terms $\check{U}^\mathsf{T} x(t_i)$ and $\left\|\left(I_n - \check{U}\check{U}^\mathsf{T}\right)x(t_i)\right\|$ can be computed before the optimization. The computation of the norm of the residual in (5.12e) is also low-dimensional as discussed in Section 2.2. Altogether, the computational complexity of the optimization problem is independent of the dimension of the preprocessed model.

If the order of the preprocessed model is reduced significantly, we expect that the computational efficiency is increased by replacing the preprocessed model with the reduced model and the scalar error bounding system. To have the advantage of increased computational efficiency while guaranteeing constraint satisfaction, the constraints are tightened by the error bound in (5.12c), which results in a loss of performance and in general in a smaller region of attraction. Thus, there is a trade-off between computational efficiency and conservatism for Δ-MPC. In comparison to $\mathcal{R}$-MPC, the Δ-MPC scheme guarantees constraint satisfaction and asymptotic stability of the closed loop, which is proven in Section 5.6. Beforehand, we present a possibility how the approximation of the cost functional due to the model reduction can be overcome.

5.5 Eliminating the Model Reduction Error in the Cost Functional

In Section 5.2, the cost functional for the preprocessed model is approximated since the state of the preprocessed model is replaced with the state estimated by the reduced model. In this section, we allow to minimize the finite horizon cost corresponding to the cost functional of the preprocessed model although a reduced model is used for the prediction if the LQR is used for the prestabilization. Hence, for this case, the approximation of the cost functional can be overcome for the proposed MPC scheme.

To eliminate the model reduction error in the cost functional, we introduce the following lemma (see [Chisci et al., 2001; Kouvaritakis et al., 2002] for similar statements for discrete-time systems).

Lemma 5.11 (Equivalent cost functional for prestabilization with the LQR). *Consider the plant model* (5.1) *and the cost functional*

$$J_{\mathrm{P}}^{\mathrm{inf}}(x_{\mathrm{P},0}, u_{\mathrm{P}}) = \int_0^\infty x_{\mathrm{P}}^{\mathsf{T}}(t) Q_{\mathrm{P}} x_{\mathrm{P}}(t) + u_{\mathrm{P}}^{\mathsf{T}}(t) R_{\mathrm{P}} u_{\mathrm{P}}(t)\,\mathrm{d}t$$

satisfying Assumption 5.1 and 5.3. Furthermore, consider the unique solution $P_{\mathrm{P}}^{\mathrm{ARE}} \succ 0$ *of the ARE*

$$P_{\mathrm{P}}^{\mathrm{ARE}} A_{\mathrm{P}} + A_{\mathrm{P}}^{\mathsf{T}} P_{\mathrm{P}}^{\mathrm{ARE}} + Q_{\mathrm{P}} - P_{\mathrm{P}}^{\mathrm{ARE}} B_{\mathrm{P}} R_{\mathrm{P}}^{-1} B_{\mathrm{P}}^{\mathsf{T}} P_{\mathrm{P}}^{\mathrm{ARE}} = 0_{n \times n} \tag{5.15}$$

and the input trajectory $u_{\mathrm{P}}(t) := u(t) - R_{\mathrm{P}}^{-1} B_{\mathrm{P}}^{\mathsf{T}} P_{\mathrm{P}}^{\mathrm{ARE}} x_{\mathrm{P}}(t)$*, which is the sum of any input trajectory* $u(t) \in \mathcal{L}_2^{n_{\mathrm{U}}}$ *and the* LQR *feedback* $-R_{\mathrm{P}}^{-1} B_{\mathrm{P}}^{\mathsf{T}} P_{\mathrm{P}}^{\mathrm{ARE}} x_{\mathrm{P}}(t)$*. Then,*

$$J_{\mathrm{P}}^{\mathrm{inf}}(x_{\mathrm{P},0}, u_{\mathrm{P}}) = x_{\mathrm{P},0}^{\mathsf{T}} P_{\mathrm{P}}^{\mathrm{ARE}} x_{\mathrm{P},0} + \int_0^\infty u^{\mathsf{T}}(t) R_{\mathrm{P}} u(t)\,\mathrm{d}t\,.$$

Proof. From Assumption 5.1 and 5.3 follows that the solution of the ARE is unique and positive definite.

The cost functional satisfies

$$\begin{aligned}
&J_{\mathrm{P}}^{\mathrm{inf}}(x_{\mathrm{P},0}, u_{\mathrm{P}}) \\
&\quad \overset{\mathrm{ARE}}{=} \int_0^\infty x_{\mathrm{P}}^{\mathsf{T}}(t)\Big(-P_{\mathrm{P}}^{\mathrm{ARE}} A_{\mathrm{P}} - A_{\mathrm{P}}^{\mathsf{T}} P_{\mathrm{P}}^{\mathrm{ARE}} + P_{\mathrm{P}}^{\mathrm{ARE}} B_{\mathrm{P}} R_{\mathrm{P}}^{-1} B_{\mathrm{P}}^{\mathsf{T}} P_{\mathrm{P}}^{\mathrm{ARE}}\Big) x_{\mathrm{P}}(t) \\
&\qquad + u_{\mathrm{P}}^{\mathsf{T}}(t) R_{\mathrm{P}} u_{\mathrm{P}}(t)\,\mathrm{d}t \\
&\quad = \int_0^\infty -\big(A_{\mathrm{P}} x_{\mathrm{P}}(t) + B_{\mathrm{P}} u_{\mathrm{P}}(t)\big)^{\mathsf{T}} P_{\mathrm{P}}^{\mathrm{ARE}} x_{\mathrm{P}}(t) - x_{\mathrm{P}}^{\mathsf{T}}(t) P_{\mathrm{P}}^{\mathrm{ARE}} \big(A_{\mathrm{P}} x_{\mathrm{P}}(t) + B_{\mathrm{P}} u_{\mathrm{P}}(t)\big) \\
&\qquad + \Big(u_{\mathrm{P}}(t) + R_{\mathrm{P}}^{-1} B_{\mathrm{P}}^{\mathsf{T}} P_{\mathrm{P}}^{\mathrm{ARE}} x_{\mathrm{P}}(t)\Big)^{\mathsf{T}} R_{\mathrm{P}} \Big(u_{\mathrm{P}}(t) + R_{\mathrm{P}}^{-1} B_{\mathrm{P}}^{\mathsf{T}} P_{\mathrm{P}}^{\mathrm{ARE}} x_{\mathrm{P}}(t)\Big)\,\mathrm{d}t \\
&\quad \overset{\Sigma_{\mathrm{P}}}{=} \int_0^\infty -\dot{x}_{\mathrm{P}}(t)^{\mathsf{T}} P_{\mathrm{P}}^{\mathrm{ARE}} x_{\mathrm{P}}(t) - x_{\mathrm{P}}^{\mathsf{T}}(t) P_{\mathrm{P}}^{\mathrm{ARE}} \dot{x}_{\mathrm{P}}(t) dt + \int_0^\infty u^{\mathsf{T}}(t) R_{\mathrm{P}} u(t)\,\mathrm{d}t \\
&\quad = x_{\mathrm{P},0}^{\mathsf{T}} P_{\mathrm{P}}^{\mathrm{ARE}} x_{\mathrm{P},0} - \lim_{t\to\infty} x_{\mathrm{P}}^{\mathsf{T}}(t) P_{\mathrm{P}}^{\mathrm{ARE}} x_{\mathrm{P}}(t) + \int_0^\infty u^{\mathsf{T}}(t) R_{\mathrm{P}} u(t)\,\mathrm{d}t\,.
\end{aligned}$$

The integrals exist since $u(t) \in \mathcal{L}_2^{n_\mathrm{U}}$ and the closed loop of the plant with the LQR is exponentially stable. From the same argument follows that $\lim_{t\to\infty} x_\mathrm{P}(t) = 0$ [Desoer and Vidyasagar, 2009, page 59], which concludes the proof. □

The input $u(t)$ in the lemma equals the input of the preprocessed model when using the LQR for prestabilization. Hence, for the preprocessed model the cost functional

$$x_0^\mathsf{T} T_\mathrm{P}^{-\mathsf{T}} P_\mathrm{P}^\mathrm{ARE} T_\mathrm{P}^{-1} x_0 + \int_0^\infty u^\mathsf{T}(t) R u(t)\,\mathrm{d}t \tag{5.16}$$

can be used. The lemma demonstrates that for the considered case the cost functional of the preprocessed model can be decomposed into a summand depending only on the initial condition and a summand that is the integral of a stage cost depending only on the input.

To discuss the consequences of Lemma 5.11 for the optimal control problems, we define that two optimization problems are equivalent, if from an optimal solution of one problem an optimal solution of the other problem can be directly computed, and the other way around [Boyd and Vandenberghe, 2004].

Consider the infinite horizon optimal control problem of $\mathcal{P}$-MPC for the preprocessed model with the cost functional in (5.16) motivated by Lemma 5.11. The first summand is independent of the solution of the optimal control problem. Hence, the first summand does not change the optimal solution. Thus, omitting the first summand results in an equivalent optimization problem. This motivates to introduce

$$\check{F}\big(x(t), u(t)\big) := u^\mathsf{T}(t) R u(t) \tag{5.17}$$

which is denoted equivalent stage cost.

In the remainder of this subsection, we assume that the LQR is used for the prestabilization.

Before discussing the implications of Lemma 5.11 for Δ-MPC, we show that the optimal control problems from MPC using the plant model and MPC using the preprocessed model are equivalent also for a finite-horizon under mild assumptions on the terminal cost.

The optimal control problem of MPC using the plant model is limited to a quadratic terminal cost that is determined by the solution of the ARE in (5.15).

Problem 5.12 (Optimization problem for MPC using the plant model).

$$\begin{aligned}
\underset{\bar{u}_{\mathrm{P}} \in \mathcal{PC}^{n_{\mathrm{U}}}_{[t_i, t_i+T]}, \bar{x}_{\mathrm{P}}}{\text{minimize}} \quad & J_{\mathcal{P}}\big(x_{\mathrm{P}}(t_i), \bar{u}_{\mathrm{P}}(\cdot; t_i)\big) = \\
& \int_{t_i}^{t_i+T} \bar{x}_{\mathrm{P}}^{\mathsf{T}}(t;t_i) Q_{\mathrm{P}} \bar{x}_{\mathrm{P}}(t;t_i) + \bar{u}_{\mathrm{P}}^{\mathsf{T}}(t;t_i) R_{\mathrm{P}} \bar{u}_{\mathrm{P}}(t;t_i) \, \mathrm{d}t \\
& \qquad + \bar{x}_{\mathrm{P}}^{\mathsf{T}}(t_i+T;t_i) P_{\mathrm{P}}^{\mathrm{ARE}} \bar{x}_{\mathrm{P}}(t_i+T;t_i) \\
\text{subject to} \quad & \dot{\bar{x}}_{\mathrm{P}}(t;t_i) = A_{\mathrm{P}} \bar{x}_{\mathrm{P}}(t;t_i) + B_{\mathrm{P}} \bar{u}_{\mathrm{P}}(t;t_i) \,, \\
& \bar{x}_{\mathrm{P}}(t_i;t_i) = x_{\mathrm{P}}(t_i) \,, \\
& C_{\mathrm{P}} \begin{bmatrix} \bar{x}_{\mathrm{P}}(t;t_i) \\ \bar{u}_{\mathrm{P}}(t;t_i) \end{bmatrix} \le d_{\mathrm{P}} \,, \\
& \bar{x}_{\mathrm{P}}(t_i+T;t_i) \in \Omega_{\mathcal{P}} \,, \\
& \text{for all } t \in [t_i, t_i+T] \,.
\end{aligned}$$

The optimal control problem of MPC using the preprocessed model and the equivalent stage cost uses no terminal cost.

Problem 5.13 (Optimization problem for MPC using the preprocessed model and the equivalent stage cost).

$$\begin{aligned}
\underset{\bar{u}_{\mathrm{P}} \in \mathcal{PC}^{n_{\mathrm{U}}}_{[t_i, t_i+T]}, \bar{x}}{\text{minimize}} \quad & \check{J}_{\mathrm{PP}}\big(T_{\mathrm{P}} x_{\mathrm{P}}(t_i), \bar{u}(\cdot; t_i)\big) := \int_{t_i}^{t_i+T} \bar{u}^{\mathsf{T}}(t;t_i) R \bar{u}(t;t_i) \, \mathrm{d}t \\
\text{subject to} \quad & \dot{\bar{x}}(t;t_i) = A \bar{x}(t;t_i) + B \bar{u}(t;t_i) \,, \\
& \bar{x}(t_i;t_i) = T_{\mathrm{P}} x_{\mathrm{P}}(t_i) \,, \\
& C \begin{bmatrix} \bar{x}(t;t_i) \\ \bar{u}(t;t_i) \end{bmatrix} \le d \,, \\
& T_{\mathrm{P}}^{-1} \bar{x}(t_i+T;t_i) \in \Omega_{\mathcal{P}} \,, \\
& \text{for all } t \in [t_i, t_i+T] \,.
\end{aligned}$$

As introduced in Section 5.2, we have $A = T_{\mathrm{P}}(A_{\mathrm{P}} - B_{\mathrm{P}} K_{\mathrm{P}}) T_{\mathrm{P}}^{-1}$, $B = T_{\mathrm{P}} B_{\mathrm{P}}$, $C = C_{\mathrm{P}} \begin{bmatrix} T_{\mathrm{P}}^{-1} & 0_{n \times n_{\mathrm{U}}} \\ -K_{\mathrm{P}} T_{\mathrm{P}}^{-1} & I_{n_{\mathrm{U}}} \end{bmatrix}$, $d = d_{\mathrm{P}}$, and $R_{\mathrm{P}} = R$ with $K_{\mathrm{P}} = K_{\mathrm{LQR}}$.

The equivalence of the two optimal control problems is stated in the following lemma.

Lemma 5.14 (Equivalence of finite-horizon optimal control problems for MPC using the plant model and using the preprocessed model). *Consider Problem 5.12 with a quadratic terminal cost determined by the solution of the ARE in* (5.15). *Additionally, consider Problem 5.13 with the equivalent cost functional and no terminal cost. Suppose that Assumption 5.1 and 5.3 are satisfied and that the preprocessed model is derived using the* LQR *and any* $T_{\mathrm{P}} \succ 0$. *Then, Problem 5.12 and Problem 5.13 are equivalent.*

Proof. From the preprocessing follows:

F1) The solution of the ARE for the preprocessed model is $P_{\mathrm{PP}}^{\mathrm{ARE}} = T_{\mathrm{P}}^{-\mathsf{T}} P_{\mathrm{P}}^{\mathrm{ARE}} T_{\mathrm{P}}^{-1}$.

F2) The LQR for the preprocessed model is 0.

Applying the prestabilization with the LQR and the state transformation of the preprocessing to Problem 5.12 results in an equivalent optimization problem. This optimization problem has the constraints of Problem 5.13 and the cost functional

$$\begin{aligned} J_{\mathrm{PP}}\big(T_{\mathrm{P}} x_{\mathrm{P}}(t_i), \bar{u}(\cdot; t_i)\big) \\ := \int_{t_i}^{t_i+T} \bar{x}^{\mathsf{T}}(t;t_i) Q \bar{x}(t;t_i) + 2\bar{x}^{\mathsf{T}}(t;t_i) S \bar{u}(t;t_i) + \bar{u}^{\mathsf{T}}(t;t_i) R \bar{u}(t;t_i)\, \mathrm{d}t \\ + \bar{x}^{\mathsf{T}}(t_i+T;t_i) T_{\mathrm{P}}^{-\mathsf{T}} P_{\mathrm{P}}^{\mathrm{ARE}} T_{\mathrm{P}}^{-1} \bar{x}(t_i+T;t_i) \end{aligned} \tag{5.20}$$

in which $Q = T_{\mathrm{P}}^{-\mathsf{T}} Q_{\mathrm{P}} T_{\mathrm{P}}^{-1} + T_{\mathrm{P}}^{-\mathsf{T}} K_{\mathrm{P}}^{\mathsf{T}} R_{\mathrm{P}} K_{\mathrm{P}} T_{\mathrm{P}}^{-1}$ and $S = -T_{\mathrm{P}}^{-\mathsf{T}} K_{\mathrm{P}}^{\mathsf{T}} R_{\mathrm{P}}$. Consider the extension of the predicted input trajectory with zero until infinity

$$\tilde{u}(t) := \begin{cases} \bar{u}(t;t_i) & \text{if } t \in [t_{i+1}\,,\, t_i + T)\,, \\ 0 & \text{if } t \in [t_i + T,\, \infty]\,. \end{cases}$$

From F2) follows that the input for $t \in [t_i + T,\, \infty]$ satisfies the control law of the LQR. Hence, with F1) follows that the terminal cost in (5.20) equals the integral of the stage cost for $t \in [t_i + T,\, \infty]$. Thus, we have

$$J_{\mathrm{PP}}\big(T_{\mathrm{P}} x_{\mathrm{P}}(t_i), \tilde{u}(\cdot)\big) = \int_{t_i}^{\infty} \tilde{x}^{\mathsf{T}}(t;t_i) Q \tilde{x}(t;t_i) + 2\tilde{x}^{\mathsf{T}}(t;t_i) S \tilde{u}(t) + \tilde{u}^{\mathsf{T}}(t) R \tilde{u}(t)\, \mathrm{d}t\,.$$

From Assumption 5.1 and 5.3 follows that the LQR stabilizes the plant model. Hence, applying $\tilde{u}(t)$ to the preprocessed model results in $\lim_{t\to\infty} x(t) = 0$. Thus, $\lim_{t\to\infty} x_{\mathrm{P}}(t) = 0$. Consequently, from Lemma 5.11 follows

$$J_{\mathrm{PP}}\big(T_{\mathrm{P}} x_{\mathrm{P}}(t_i), \tilde{u}(\cdot)\big) = x_{\mathrm{P}}^{\mathsf{T}}(t_i) P_{\mathrm{P}}^{\mathrm{ARE}} x_{\mathrm{P}}(t_i) + \int_{t_i}^{\infty} \tilde{u}^{\mathsf{T}}(t) R \tilde{u}(t)\, \mathrm{d}t\,.$$

Since $\tilde{u}(t) = 0$ for $t \in [t_i + T,\, \infty]$, we have

$$\begin{aligned} J_{\mathrm{PP}}\big(T_{\mathrm{P}} x_{\mathrm{P}}(t_i), \bar{u}(\cdot;t_i)\big) &= x_{\mathrm{P}}^{\mathsf{T}}(t_i) P_{\mathrm{P}}^{\mathrm{ARE}} x_{\mathrm{P}}(t_i) + \int_{t_i}^{t_i+T} \bar{u}^{\mathsf{T}}(t;t_i) R \bar{u}(t;t_i)\, \mathrm{d}t \\ &= x_{\mathrm{P}}^{\mathsf{T}}(t_i) P_{\mathrm{P}}^{\mathrm{ARE}} x_{\mathrm{P}}(t_i) + \check{J}_{\mathrm{PP}}\big(T_{\mathrm{P}} x_{\mathrm{P}}(t_i), \bar{u}(\cdot;t_i)\big)\,. \end{aligned}$$

Altogether, Problem 5.12 can be transformed into an optimization problem that differs from Problem 5.13 only by a summand in the cost functional. Since this summand depends only on the initial condition, Problem 5.12 and Problem 5.13 are equivalent. □

The lemma establishes the equivalence of MPC using the plant model and MPC using the preprocessed model with the equivalent stage cost for the considered design parameters.

Consider the Δ-MPC scheme for the preprocessed model with the equivalent stage cost in (5.17) and no terminal cost in addition to the case of Lemma 5.14. Then, the following conclusions for the Δ-MPC scheme are valid.

1. The equivalent stage cost is independent of the state of the preprocessed model. Hence, *using the reduced model for the prediction introduces no error in the cost functional* – both in the infinite horizon case as shown in Lemma 5.11 and the finite horizon case as shown in Lemma 5.14.

2. The cost functional of Δ-MPC is identical to the cost function of MPC using the preprocessed model in Problem 5.13. Since Problem 5.13 is equivalent to Problem 5.12, the Δ-MPC scheme implicitly *minimizes the quasi-infinite horizon cost functional of* $\mathcal{P}$-MPC.

3. Nevertheless, the Δ-MPC scheme, in general, leads to a worse performance compared to $\mathcal{P}$-MPC since the state and input constraints are tightened by the error bound. Suppose that the assumptions of Theorem 5.22 are satisfied. Then, the infinite horizon cost functional value of Δ-MPC is bounded by the cost functional value of the optimal solution at the initial time instant as shown below in (5.35), i.e., $\int_0^\infty u_\Delta^\mathsf{T}(t) R u_\Delta(t)\,\mathrm{d}t \leq J_\Delta^*\big(x_\mathrm{R}^*(0), u_\Delta^*(\cdot;0)\big)$. Together with Lemma 5.11 follows $J_\mathrm{P}^\mathrm{inf}(x_{\mathrm{P},0}, u_\mathrm{P}) \leq x_{\mathrm{P},0}^\mathsf{T} P_\mathrm{P}^\mathrm{ARE} x_{\mathrm{P},0} + J_\Delta^*\big(x_\mathrm{R}^*(0), u_\Delta^*(\cdot;0)\big)$. Consequently, the Δ-MPC scheme *guarantees an upper bound for the cost functional value of the closed loop with the plant.*

4. The model predictive controller deviates the input of the plant from the optimal solution of the unconstrained case, i.e, the LQR control law $u_\mathrm{P}(t) = u(t) - R_\mathrm{P}^{-1} B_\mathrm{P}^\mathsf{T} P_\mathrm{P}^\mathrm{ARE} x_\mathrm{P}(t)$, in order to satisfy the constraints. If no state, input, or terminal constraint is active in Problem 5.8, then the optimal solution is $u(t) \equiv 0$, i.e., the LQR is optimal without constraints. Hence, if the equilibrium $(x_\mathrm{R}, \Delta) = 0$ with $u = 0$ is contained strictly within the terminal constraint, then the proposed MPC scheme is *locally optimal.*

The initial condition of the reduced model is used in Problem 5.8 as decision variable. This is used by the model predictive controller to minimize the deviation from the LQR control law. For this reason, we allow for the origin beside the projection of the state of the preprocessed model and the predicted state of the reduced model as possible initial condition in Problem 5.8.

The equivalent stage cost (5.17) results in $Q = 0_{n\times n}$ and $S = 0_{n\times n_{\mathrm{U}}}$. Hence, $Q - SR^{-1}S^{\mathsf{T}} = 0_{n\times n}$ is not positive definite. Thus, Assumption 5.6 is not satisfied. Relaxing Assumption 5.6 to $Q - SR^{-1}S^{\mathsf{T}}$ is symmetric and positive semidefinite and R is symmetric and positive definite prohibits proving asymptotic stability of the preprocessed model controlled by Δ-MPC, since the origin is not necessarily an equilibrium of the closed-loop system. To illustrate this, consider $Q = S = R = 1$, $A = -1$, $B = 1$, $V = W = 1$, $E_\Delta(x_{\mathrm{R}}) = 0$, $x(t_i) = 0$, $\bar{x}_{\mathrm{R}}(t_i; t_{i-1}) = 1$, and the solution of Problem 5.8 given by $x_{\mathrm{R}}(t_i) = 1$ and $\bar{u}(t; t_i) = -\bar{x}_{\mathrm{R}}(t; t_i)$. For large enough state and input constraints of the preprocessed model, this solution is feasible. Since $F_{\mathrm{R}}(\bar{x}_{\mathrm{R}}(t; t_i), -\bar{x}_{\mathrm{R}}(t; t_i)) = \bar{x}_{\mathrm{R}}^{\mathsf{T}}(t; t_i)(Q_{\mathrm{R}} - 2S_{\mathrm{R}} + R)\bar{x}_{\mathrm{R}}(t; t_i) = 0$ results in $J_\Delta\big(1, -\bar{x}_{\mathrm{R}}(t; t_i)\big) = 0$, this solution is optimal. Applying $\bar{u}(t; t_i) = -\bar{x}_{\mathrm{R}}(t; t_i)$ to the preprocessed model results in $\dot{x}(t) \neq 0$. Thus, the origin is not necessarily an equilibrium of the closed-loop system.

Consequently, we replace Assumption 5.6 with the following relaxed assumption, which excludes the case $S \neq 0_{n\times n_{\mathrm{U}}}$ if $Q - SR^{-1}S^{\mathsf{T}} \succeq 0$ but not $Q - SR^{-1}S^{\mathsf{T}} \succ 0$.

Assumption 5.15. *The matrix R is symmetric and positive definite. Furthermore, the matrices Q and S satisfy at least one of the following statements:*

i) The matrix $Q - SR^{-1}S^{\mathsf{T}}$ is symmetric and positive definite.

ii) The matrix Q is symmetric and positive semidefinite and $S = 0_{n\times n_{\mathrm{U}}}$.

In the Δ-MPC scheme, the matrices $Q_{\mathrm{R}} = V^{\mathsf{T}}QV$ and $S_{\mathrm{R}} = V^{\mathsf{T}}S$ and not Q and S appear in the optimization problem. In the proof of asymptotic stability, we use the following assumption on Q_{R}, S_{R}, and R.

Assumption 5.16. *The matrix R is symmetric and positive definite. Furthermore, the matrices Q_{R} and S_{R} satisfy at least one of the following statements:*

i) The matrix $Q_{\mathrm{R}} - S_{\mathrm{R}}R^{-1}S_{\mathrm{R}}^{\mathsf{T}}$ is symmetric and positive definite.

ii) The matrix Q_{R} is symmetric and positive semidefinite and $S_{\mathrm{R}} = 0_{n_{\mathrm{R}}\times n_{\mathrm{U}}}$.

For model reduction by projection, the matrices $V, W \in \mathbb{R}^{n\times n_{\mathrm{R}}}$ satisfy $W^{\mathsf{T}}V = I_{n_{\mathrm{R}}}$ and $n_{\mathrm{R}} < n$. Hence, the matrix V has full rank. Consequently, Assumption 5.15 is sufficient but not necessary for Assumption 5.16.

In this section, we have shown that an equivalent cost functional allows to eliminate the influence of the model reduction error on the cost functional when using the LQR for prestabilization of the plant. To allow for the equivalent cost functional, we have generalized the assumption on the cost functional.

5.6 Guaranteeing Asymptotic Stability

In this section, we prove asymptotic stability of the preprocessed model in closed loop with Δ-MPC using the generalized Assumption 5.16. At the beginning, we define the terminal set based on a terminal controller such that, first, the terminal set is positive invariant for the reduced model under the terminal controller and, second, the constraints are satisfied within the terminal set when using this terminal controller. These properties will be used afterwards to show recursive feasibility. Furthermore, we define the terminal cost, which is used at the end of this section to prove asymptotic stability of the plant controlled by Δ-MPC.

To prove invariance of the terminal set, the following assumptions are used below.

Assumption 5.17. *The error bounding system* (5.8) *is asymptotically stable.*

Assumption 5.18. *The pair* $(A_{\mathrm{R}}, B_{\mathrm{R}})$ *is stabilizable.*

Assumption 5.19. *The pair* $\left(A_{\mathrm{R}} - B_{\mathrm{R}} R^{-1} S_{\mathrm{R}}^{\mathsf{T}}, (Q_{\mathrm{R}} - S_{\mathrm{R}} R^{-1} S_{\mathrm{R}}^{\mathsf{T}})^{1/2}\right)$ *is detectable.*

If the preprocessed model is stabilizable, an asymptotically stable error bounding system can always be achieved by a prestabilization of the plant as shown in Section 4.3. If asymptotic stability is preserved by the model reduction, Assumption 5.18 and 5.19 are implicitly satisfied by Assumption 5.16 and 5.17, since Assumption 5.17 requires an asymptotically stable preprocessed model and, hence, by Assumption 5.16 either $(Q_{\mathrm{R}} - S_{\mathrm{R}} R^{-1} S_{\mathrm{R}}^{\mathsf{T}})^{1/2}$ has full rank or $A_{\mathrm{R}} - B_{\mathrm{R}} R^{-1} S_{\mathrm{R}}^{\mathsf{T}} = A_{\mathrm{R}}$ is Hurwitz. Since the matrix $Q_{\mathrm{R}} - S_{\mathrm{R}} R^{-1} S_{\mathrm{R}}^{\mathsf{T}}$ is positive semidefinite, the unique and symmetric square root in Assumption 5.19 exists.

The following lemma states a possible choice for the terminal set and terminal cost (to compare with nominal MPC see, e.g., Assumption 2.11 or [Chen and Allgöwer, 1998, Lemma 1 (b)]).

Lemma 5.20 (Terminal set and terminal cost)**.** *Consider the preprocessed model* (5.4)*, the reduced model* (5.7) *defined by* $V, W \in \mathbb{R}^{n \times n_{\mathrm{R}}}$ *in which* $n_{\mathrm{R}} < n$*, and the error bounding system* (5.8)*. Suppose that Assumptions 5.5, 5.16–5.19 are satisfied. Then, there exist* $\gamma_{\mathrm{x}} > 0$*,* $\gamma_{\Delta} > 0$*,* $P_{\Delta}^{\Omega} \succ 0$*,* $Q_{\Delta}^{\Omega} \succeq 0$*, and* K_{Ω} *that determine the terminal set*

$$\Omega_{\Delta} := \left\{ (x_{\mathrm{R}}, \Delta) \in \mathbb{R}^{n_{\mathrm{R}}} \times \mathbb{R} \;\middle|\; x_{\mathrm{R}}^{\mathsf{T}} P_{\Delta}^{\Omega} x_{\mathrm{R}} \leq \gamma_{\mathrm{x}},\, 0 \leq \Delta \leq \gamma_{\Delta} \right\}, \tag{5.21}$$

and terminal cost $E_{\Delta}(x_{\mathrm{R}}) := x_{\mathrm{R}}^{\mathsf{T}} Q_{\Delta}^{\Omega} x_{\mathrm{R}}$ *such that*

i) the terminal set Ω_{Δ} *is positive invariant for the interconnection of the reduced model and the error bounding system under the control law* $u(t) = -K_{\Omega} x_{\mathrm{R}}(t)$*,*

ii) for all $(x_{\mathrm{R}}, \Delta) \in \Omega_{\Delta}$ *the state and input constraints* (5.12c) *are satisfied for* $u(t) = -K_{\Omega} x_{\mathrm{R}}(t)$*, and*

iii) for all $(x_{\mathrm{R}}, \Delta) \in \Omega_{\Delta}$ *the closed loop with the reduced model and the control law* $u(t) = -K_{\Omega} x_{\mathrm{R}}(t)$ *satisfies*

$$\dot{E}_{\Delta}(x_{\mathrm{R}}) + F_{\mathrm{R}}(x_{\mathrm{R}}, -K_{\Omega} x_{\mathrm{R}}) \leq 0\,. \tag{5.22}$$

Proof. Consider the new input $\tilde{u}(t) := u(t) + R^{-1} S_{\mathrm{R}}^{\mathsf{T}} x_{\mathrm{R}}(t)$, which results in the reduced model with the dynamic matrix $\tilde{A}_{\mathrm{R}} = A_{\mathrm{R}} - B_{\mathrm{R}} R^{-1} S_{\mathrm{R}}^{\mathsf{T}}$ and the stage cost $\tilde{F}_{\mathrm{R}}\big(x_{\mathrm{R}}(t), \tilde{u}(t)\big) = x_{\mathrm{R}}^{\mathsf{T}}(t)\tilde{Q}_{\mathrm{R}}x_{\mathrm{R}}(t) + \tilde{u}^{\mathsf{T}}(t)R\tilde{u}(t)$ in which $\tilde{Q}_{\mathrm{R}} = Q_{\mathrm{R}} - S_{\mathrm{R}}R^{-1}S_{\mathrm{R}}^{\mathsf{T}}$. From Assumption 5.18 follows that the pair $(\tilde{A}_{\mathrm{R}}, B_{\mathrm{R}})$ is stabilizable. From Assumption 5.16 follows that $\tilde{Q}_{\mathrm{R}}$ is symmetric and positive semidefinite. Furthermore, Assumption 5.19 guarantees that the pair $\big(\tilde{A}_{\mathrm{R}}, \tilde{Q}_{\mathrm{R}}^{1/2}\big)$ is detectable. Altogether, there exists a unique solution $P_{\mathrm{R}}^{\mathrm{ARE}} \succeq 0$ of the ARE

$$P_{\mathrm{R}}^{\mathrm{ARE}}\tilde{A}_{\mathrm{R}} + \tilde{A}_{\mathrm{R}}^{\mathsf{T}}P_{\mathrm{R}}^{\mathrm{ARE}} + \tilde{Q}_{\mathrm{R}} - P_{\mathrm{R}}^{\mathrm{ARE}}B_{\mathrm{R}}R^{-1}B_{\mathrm{R}}^{\mathsf{T}}P_{\mathrm{R}}^{\mathrm{ARE}} = 0_{n_{\mathrm{R}}\times n_{\mathrm{R}}} \tag{5.23}$$

and $A_{\mathrm{K}} = A_{\mathrm{R}} - B_{\mathrm{R}}R^{-1}\big(B_{\mathrm{R}}^{\mathsf{T}}P_{\mathrm{R}}^{\mathrm{ARE}} + S_{\mathrm{R}}^{\mathsf{T}}\big)$ is Hurwitz [Kwakernaak and Sivan, 1972].

For the proof of Lemma 5.20, consider the LQR as terminal controller $K_{\Omega} := R^{-1}\big(B_{\mathrm{R}}^{\mathsf{T}}P_{\mathrm{R}}^{\mathrm{ARE}} + S_{\mathrm{R}}^{\mathsf{T}}\big)$ and for the terminal cost $Q_{\Delta}^{\Omega} := P_{\mathrm{R}}^{\mathrm{ARE}}$. The terminal region for the reduced state is determined by $P_{\Delta}^{\Omega} := P_{\mathrm{R}}^{\mathrm{L}}$ in which $P_{\mathrm{R}}^{\mathrm{L}}$ is the unique positive definite solution of the Lyapunov equation

$$P_{\mathrm{R}}^{\mathrm{L}}A_{\mathrm{K}} + A_{\mathrm{K}}^{\mathsf{T}}P_{\mathrm{R}}^{\mathrm{L}} + Q_{\mathrm{R}} + S_{\mathrm{R}}K_{\Omega} + K_{\Omega}^{\mathsf{T}}S_{\mathrm{R}}^{\mathsf{T}} + K_{\Omega}^{\mathsf{T}}RK_{\Omega} + \kappa I_{n_{\mathrm{R}}} = 0_{n_{\mathrm{R}}\times n_{\mathrm{R}}}\,, \tag{5.24}$$

in which $\kappa \geq 0$ is chosen such that $Q_{\mathrm{R}} + S_{\mathrm{R}}K_{\Omega} + K_{\Omega}^{\mathsf{T}}S_{\mathrm{R}}^{\mathsf{T}} + K_{\Omega}^{\mathsf{T}}RK_{\Omega} + \kappa I_{n_{\mathrm{R}}}$ is positive definite.

i) The reduced model is independent of the error bound $\Delta(t)$. Since P_{Δ}^{Ω} solves the Lyapunov equation (5.24), the set $\Omega_{\mathrm{x}} := \big\{x_{\mathrm{R}} \in \mathbb{R}^{n_{\mathrm{R}}} \;\big|\; x_{\mathrm{R}}^{\mathsf{T}}P_{\Delta}^{\Omega}x_{\mathrm{R}} \leq \gamma_{\mathrm{x}}\big\}$ is positive invariant for the closed loop of the reduced model and the terminal controller.

Furthermore, for all $x_{\mathrm{R}}(t) \in \Omega_{\mathrm{x}}$,

$$\begin{aligned}
\dot{\Delta}(t) &\overset{(5.8)}{=} -\beta\Delta(t) + \alpha\left\|\big(I_n - VW^{\mathsf{T}}\big)\big(AV - BK_{\Omega}\big)x_{\mathrm{R}}(t)\right\| \\
&\leq -\beta\Delta(t) + \alpha\left\|\big(I_n - VW^{\mathsf{T}}\big)\big(AV - BK_{\Omega}\big)\Big(P_{\Delta}^{\Omega}\Big)^{-1/2}\right\|\left\|\Big(P_{\Delta}^{\Omega}\Big)^{1/2}x_{\mathrm{R}}(t)\right\| \\
&\overset{x_{\mathrm{R}}(t)\in\Omega_{\mathrm{R}}}{\leq} -\beta\Delta(t) + \alpha\sqrt{\gamma_{\mathrm{x}}}\left\|\big(I_n - VW^{\mathsf{T}}\big)\big(AV - BK_{\Omega}\big)\Big(P_{\Delta}^{\Omega}\Big)^{-1/2}\right\|.
\end{aligned}$$

Choosing γ_{x} and γ_{Δ} such that

$$\alpha\sqrt{\gamma_{\mathrm{x}}}\left\|\big(I_n - VW^{\mathsf{T}}\big)\big(AV - BK_{\Omega}\big)\Big(P_{\Delta}^{\Omega}\Big)^{-1/2}\right\| \leq \beta\,\gamma_{\Delta} \tag{5.25}$$

yields

$$\dot{\Delta}(t) \leq -\beta\,\Delta(t) + \beta\,\gamma_{\Delta}\,.$$

Since $\beta > 0$ is guaranteed by Assumption 5.17, for all $\Delta(t) > \gamma_{\Delta}$, we have $\dot{\Delta}(t) < 0$ and $\Delta(t) = \gamma_{\Delta}$ results in $\dot{\Delta}(t) \leq 0$. In addition, from (5.8) follows $0 \leq \Delta(t)$. Consequently, *i)* is established.

ii) Using the triangle inequality yields

$$\begin{aligned}
c_k^{\mathsf{T}}\begin{bmatrix} V \\ -K_{\Omega}\end{bmatrix}x_{\mathrm{R}} + \Delta\left\|c_k^{\mathsf{T}}\begin{bmatrix} I_n \\ 0_{n_{\mathrm{U}}\times n}\end{bmatrix}\right\| &\leq \left|c_k^{\mathsf{T}}\begin{bmatrix} V \\ -K_{\Omega}\end{bmatrix}x_{\mathrm{R}} + \Delta\left\|c_k^{\mathsf{T}}\begin{bmatrix} I_n \\ 0_{n_{\mathrm{U}}\times n}\end{bmatrix}\right\|\right| \\
&\leq \left\|c_k^{\mathsf{T}}\begin{bmatrix} V \\ -K_{\Omega}\end{bmatrix}\Big(P_{\Delta}^{\Omega}\Big)^{-1/2}\right\|\left\|\Big(P_{\Delta}^{\Omega}\Big)^{1/2}x_{\mathrm{R}}\right\| + \Delta\left\|c_k^{\mathsf{T}}\begin{bmatrix} I_n \\ 0_{n_{\mathrm{U}}\times n}\end{bmatrix}\right\|, \quad k = 1,\ldots,n_{\mathrm{C}}\,.
\end{aligned}$$

Therefore, the constraints (5.12c) are satisfied for all $(x_{\mathrm{R}}, \Delta) \in \Omega_\Delta$ if

$$\sqrt{\gamma_{\mathrm{x}}}\left\| c_k^{\mathsf{T}} \begin{bmatrix} V \\ -K_\Omega \end{bmatrix} \left(P_\Delta^\Omega\right)^{-1/2} \right\| + \gamma_\Delta \left\| c_k^{\mathsf{T}} \begin{bmatrix} I_n \\ 0_{n_{\mathrm{U}} \times n} \end{bmatrix} \right\| \leq d_k\,, \qquad k = 1, \ldots, n_{\mathrm{C}}\,. \quad (5.26)$$

Since the equilibrium is in the interior of the constraint set, we have $d_k > 0$ for all $k = 1, \ldots, n_{\mathrm{C}}$. Thus, (5.26) determines a set for γ_{x} and γ_Δ containing $\gamma_{\mathrm{x}} = \gamma_\Delta = 0$ in the interior. Within this set, (5.25) can always be satisfied by choosing γ_{x} small enough. Altogether, parameters $\gamma_{\mathrm{x}} > 0$ and $\gamma_\Delta > 0$ satisfying both, (5.25) and (5.26), always exist. Consequently, *ii)* follows by choosing $\gamma_{\mathrm{x}} > 0$ and $\gamma_\Delta > 0$ sufficiently small.

iii) Since the terminal cost is determined by the solution of the ARE (5.23) and the LQR is used as terminal controller, with $\dot{x}_{\mathrm{R}}(t) = \left(A_{\mathrm{R}} - B_{\mathrm{R}} K_\Omega\right) x_{\mathrm{R}}(t)$ and the ARE (5.23) follows $\dot{E}_\Delta(x_{\mathrm{R}}) + F_{\mathrm{R}}(x_{\mathrm{R}}, -K_\Omega x_{\mathrm{R}}) = 0$. □

The preprocessed model in closed loop with Δ-MPC leads to recursively feasible optimization problems as stated in the following theorem (to compare with nominal MPC see, e.g., [Chen and Allgöwer, 1998, Lemma 2]).

Theorem 5.21 (Recursive feasibility of Δ-MPC). *Consider the preprocessed model* (5.4)*, the reduced model* (5.7) *defined by* $V, W \in \mathbb{R}^{n \times n_{\mathrm{R}}}$ *in which* $n_{\mathrm{R}} < n$*, and the error bounding system* (5.8)*. Suppose that*

- *the terminal set* Ω_Δ *of the form* (5.21) *and the terminal cost* $E(x_{\mathrm{R}}) = x_{\mathrm{R}}^{\mathsf{T}} Q_\Delta^\Omega x_{\mathrm{R}}$ *with* $\gamma_{\mathrm{x}} > 0$, $\gamma_\Delta > 0$, $P_\Delta^\Omega \succ 0$, $Q_\Delta^\Omega \succeq 0$, *and* K_Ω *are chosen such that i)–iii) in Lemma 5.20 are satisfied and*
- *the input of the preprocessed model is given by* $u_\Delta(\cdot)$ *defined in* (5.13).

Then, feasibility of Problem 5.8 at the sampling instant $t_0 = 0$ *implies feasibility at all subsequent sampling instants* $t_i = i\delta$, $i \in \mathbb{N}_0$.

Proof. The proof is based on standard arguments in MPC, which can be found, e.g., in [Chen and Allgöwer, 1998; Findeisen et al., 2003; Mayne et al., 2000].

In the following, we show that from feasibility at time t_i follows feasibility at time $t_{i+1} = t_i + \delta$. Then, the statement of the theorem follows by induction.

Assume that a feasible solution $x_{\mathrm{R}}(t_i)$, $\bar{u}(\cdot; t_i)$, $\bar{x}_{\mathrm{R}}(\cdot; t_i)$, $\bar{\Delta}(\cdot; t_i)$ of Problem 5.8 at time t_i is given. We consider the solution candidate at time t_{i+1} determined by $\tilde{x}_{\mathrm{R}}(t_{i+1}) := \bar{x}_{\mathrm{R}}(t_{i+1}; t_i)$, $\tilde{\Delta}(t_{i+1}) := \min\left(\alpha \| x(t_{i+1}) - V\tilde{x}_{\mathrm{R}}(t_{i+1}) \|, \bar{\Delta}(t_{i+1}; t_i)\right)$,

$$\tilde{u}(t) := \begin{cases} \bar{u}(t; t_i) & \text{if } t \in [t_{i+1}\,,\, t_i + T)\,, \\ -K_\Omega \tilde{x}_{\mathrm{R}}(t) & \text{if } t \in [t_i + T,\, t_{i+1} + T]\,, \end{cases} \quad (5.27)$$

(5.12a), and (5.12e). For all $t \in [t_{i+1}, t_i + T)$, we have $\tilde{x}_{\mathrm{R}}(t) = \bar{x}_{\mathrm{R}}(t; t_i)$, $\tilde{u}(t) = \bar{u}(t; t_i)$ and consequently

$$\dot{\tilde{\Delta}}(t) - \dot{\bar{\Delta}}(t; t_i) = -\beta\left(\tilde{\Delta}(t) - \bar{\Delta}(t; t_i)\right).$$

Thus, we conclude from $\tilde{\Delta}(t_{i+1}) \leq \bar{\Delta}(t_{i+1}; t_i)$ that $\tilde{\Delta}(t) \leq \bar{\Delta}(t; t_i)$ for all $t \in [t_{i+1}, t_i + T)$. Hence, from feasibility of $\bar{u}(\cdot)$, $\bar{x}_{\mathrm{R}}(t_i; t_i)$, $\bar{\Delta}(t_i; t_i)$, follows satisfaction of (5.12c) for $t \in [t_{i+1}\,,\ t_i + T)$. Applying $\tilde{u}(\cdot)$ drives the reduced model such that $\big(\tilde{x}_{\mathrm{R}}(t_i + T), \tilde{\Delta}(t_i + T)\big) \in \Omega_\Delta$. For $t \in [t_i + T,\ t_{i+1} + T]$, the candidate input is given by the terminal controller and thus (5.12c) and (5.12d) are satisfied according to Lemma 5.20. Furthermore, the solution candidate satisfies (5.12f). To conclude the feasibility of Problem 5.8 at time t_{i+1}, we note that the cost functional value for the candidate solution is finite. □

With the results of Theorem 5.21 and Lemma 5.20, we are ready to state the main result concerning asymptotic stability of the preprocessed model controlled with the MPC scheme using the reduced model and error bound (to compare with nominal MPC see, e.g., Theorem 2.12 or [Chen and Allgöwer, 1998, Theorem 1]).

Theorem 5.22 (Asymptotic stability of Δ-MPC). *Consider the preprocessed model* (5.4)*, the reduced model* (5.7) *defined by* $V, W \in \mathbb{R}^{n \times n_{\mathrm{R}}}$ *in which* $n_{\mathrm{R}} < n$*, and the error bounding system* (5.8)*. Suppose that*

- *Assumptions 5.5, 5.16, and 5.17 are satisfied,*
- *the terminal set* Ω_Δ *of the form* (5.21) *and the terminal cost* $E(x_{\mathrm{R}}) = x_{\mathrm{R}}^{\mathsf{T}} Q_\Delta^\Omega x_{\mathrm{R}}$ *with* $\gamma_{\mathrm{x}} > 0$, $\gamma_\Delta > 0$, $P_\Delta^\Omega \succ 0$, $Q_\Delta^\Omega \succeq 0$, *and* K_Ω *are chosen such that i)–iii) in Lemma 5.20 are fulfilled, and*
- *Problem 5.8 is feasible at the sampling instant* $t_0 = 0$.

Then,

i) the origin of the closed loop of the preprocessed model with the model predictive controller u_Δ *given by Algorithm 5.9 is asymptotically stable.*

ii) The state and input constraints (5.5) *are satisfied for the closed loop of the preprocessed model with the model predictive controller* u_Δ *given by Algorithm 5.9.*

iii) The region of attraction of the closed loop is given by the set of all initial states $x(t_0)$ *for which Problem 5.8 is feasible.*

Proof. First, we show the following three facts:

F1) There exists a finite constant $c > 0$ such that

$$\|u\|^2 \leq c\, F_{\mathrm{R}}(x_{\mathrm{R}}, u)\,. \tag{5.28}$$

F2) The $\mathcal{L}_2$-norm of the input of the preprocessed model controlled by Δ-MPC is bounded by

$$\|u_\Delta(\cdot)\|_{\mathcal{L}_2}^2 \leq c \int_{t_0}^{\infty} F_{\mathrm{R}}\big(x_\Delta(t), u_\Delta(t)\big)\,\mathrm{d}t \leq c\, J_\Delta^*\big(x_{\mathrm{R}}^*(t_0), u_\Delta^*(\cdot; t_0)\big)\,. \tag{5.29}$$

F3) For all initial conditions of the preprocessed model with $\|x(t_0)\| \leq \gamma_\Delta/\alpha$, the input of the preprocessed model controlled by Δ-MPC satisfies $u_\Delta(t) = 0$ for all $t \geq t_0$.

F1) Consider

$$F_\mathrm{R}(x_\mathrm{R}, u) = \begin{bmatrix} x_\mathrm{R}^\mathsf{T} & u^\mathsf{T} \end{bmatrix} \tilde{Q} \begin{bmatrix} x_\mathrm{R} \\ u \end{bmatrix} \quad \text{in which } \tilde{Q} := \begin{bmatrix} Q_\mathrm{R} & S_\mathrm{R} \\ S_\mathrm{R}^\mathsf{T} & R \end{bmatrix}.$$

For case *i)* of Assumption 5.16 follows using the Schur complement that $\tilde{Q}$ is positive definite. Together with $\|u\|^2 \leq \left\| \begin{bmatrix} x_\mathrm{R}^\mathsf{T} & u^\mathsf{T} \end{bmatrix} \right\|^2 \leq \sigma_{\min}^{-2}(\tilde{Q}) F_\mathrm{R}(x_\mathrm{R}, u)$ follows that inequality (5.28) is satisfied for case *i)* of Assumption 5.16 with $c = \sigma_{\min}^{-2}(\tilde{Q})$ finite. For case *ii)* of Assumption 5.16 the cross term S_R is zero and Q_R is positive semidefinite. Hence, $\sigma_{\min}^2(R)\|u\|^2 \leq u^\mathsf{T} R u \leq F_\mathrm{R}(x_\mathrm{R}, u)$. Thus, inequality (5.28) is satisfied for case *ii)* of Assumption 5.16 with $c = \sigma_{\min}^{-2}(R)$ finite. Consequently, F1) is established.

F2) We first show along the lines of [Chen and Allgöwer, 1998, proof of Lemma 3] that for all $\tau \in (t_i, t_{i+1}]$, $t_i \geq t_0$ we have

$$J_\Delta^*\big(x_\mathrm{R}^*(\tau), u_\Delta^*(\cdot;\tau)\big) - J_\Delta^*\big(x_\mathrm{R}^*(t_i), u_\Delta^*(\cdot;t_i)\big) \leq -\int_{t_i}^{\tau} F_\mathrm{R}\big(x_\Delta(t), u_\Delta(t)\big)\,\mathrm{d}t\,, \quad (5.30)$$

in which $x_\Delta(t) := x_\Delta^*(t;t_i)$ and $u_\Delta(t) := u_\Delta^*(t;t_i)$ for $t_i \leq t < t_{i+1}$.

Feasibility for all sampling instants $t_i = i\delta$, $i \in \mathbb{N}_0$ is guaranteed by feasibility at $t_0 = 0$ and Theorem 5.21.

Consider the optimal solution of Problem 5.8 at time t_i for the state of the preprocessed model $x(t_i)$ denoted by $u_\Delta^*(t;t_i)$, $x_\mathrm{R}^*(t_i)$ and the associated predicted state $x_\Delta^*(t;t_i)$ and error bound $\Delta^*(t;t_i)$ for $t \in [t_i, t_i + T]$. The optimal cost is

$$J_\Delta^*\big(x_\mathrm{R}^*(t_i), u_\Delta^*(\cdot;t_i)\big) = \int_{t_i}^{t_i+T} F_\mathrm{R}\big(x_\Delta^*(t;t_i), u_\Delta^*(t;t_i)\big)\,\mathrm{d}t + E_\Delta\big(x_\Delta^*(t_i+T;t_i)\big)\,. \quad (5.31)$$

To compute an upper bound for the cost functional $J_\Delta^*\big(x_\mathrm{R}^*(\tau), u_\Delta^*(\cdot;\tau)\big)$, consider the feasible input trajectory from the proof of Theorem 5.21

$$\tilde{u}(t) = \begin{cases} u_\Delta^*(t;t_i) & \text{if } t \in [\tau, t_i + T)\,, \\ -K_\Omega \tilde{x}_\mathrm{R}(t) & \text{if } t \in [t_i + T, \tau + T] \end{cases} \quad (5.32)$$

in which $\tilde{x}_\mathrm{R}(t)$ for $t \in [\tau, \tau + T]$ is determined by (5.12a) and $\tilde{x}_\mathrm{R}(\tau) = x_\Delta^*(\tau;t_i)$. Since $\tilde{x}_\mathrm{R}(t) = x_\Delta^*(t;t_i)$ for $t \in [\tau, t_i + T]$, the cost functional value associated with

$\tilde{x}_{\mathrm{R}}(\tau)$ and $\tilde{u}(\cdot)$ satisfies for any $\tau \in (t_i, t_{i+1}]$

$$
\begin{aligned}
J_\Delta\big(\tilde{x}_{\mathrm{R}}(\tau), \tilde{u}(\cdot)\big) &= \int_\tau^{\tau+T} F_{\mathrm{R}}\big(\tilde{x}_{\mathrm{R}}(t), \tilde{u}(t)\big)\,\mathrm{d}t + E_\Delta\big(\tilde{x}(\tau+T)\big) \\
&\overset{(5.32)}{=} \int_\tau^{t_i+T} F_{\mathrm{R}}\big(x_\Delta^*(t;t_i), u_\Delta^*(t;t_i)\big)\,\mathrm{d}t + \int_{t_i+T}^{\tau+T} F_{\mathrm{R}}\big(\tilde{x}_{\mathrm{R}}(t), -K_\Omega \tilde{x}_{\mathrm{R}}(t)\big)\,\mathrm{d}t \\
&\quad + E_\Delta\big(\tilde{x}(\tau+T)\big) \\
&\overset{(5.31)}{=} J_\Delta^*\big(x_{\mathrm{R}}^*(t_i), u_\Delta^*(\cdot;t_i)\big) - \int_{t_i}^{\tau} F_{\mathrm{R}}\big(x_\Delta^*(t;t_i), u_\Delta^*(t;t_i)\big)\,\mathrm{d}t - E_\Delta\big(x_\Delta^*(t_i+T;t_i)\big) \\
&\quad + \int_{t_i+T}^{\tau+T} F_{\mathrm{R}}\big(\tilde{x}_{\mathrm{R}}(t), -K_\Omega \tilde{x}_{\mathrm{R}}(t)\big)\,\mathrm{d}t + E_\Delta\big(\tilde{x}(\tau+T)\big)\,. \qquad (5.33)
\end{aligned}
$$

To bound the influence of the stage cost for $t \in [t_i+T, \tau+T]$, Lemma 5.20 is used. Since $\big(x_\Delta^*(t_i+T;t_i), \Delta^*(t_i+T;t_i)\big) \in \Omega_\Delta$ and the terminal set is invariant under the terminal controller, inequality (5.22) is satisfied for $t \in [t_i+T, \tau+T]$. Integrating (5.22) from t_i+T to $\tau+T$ gives

$$
E_\Delta\big(\tilde{x}(\tau+T)\big) - E_\Delta\big(\tilde{x}(t_i+T)\big) + \int_{t_i+T}^{\tau+T} F_{\mathrm{R}}\big(\tilde{x}_{\mathrm{R}}(t), -K_\Omega \tilde{x}_{\mathrm{R}}(t)\big)\,\mathrm{d}t \le 0\,. \qquad (5.34)
$$

Since $\tilde{x}_{\mathrm{R}}(t_i+T) = x_\Delta^*(t_i+T;t_i)$, combining (5.33) and (5.34) results in

$$
J_\Delta(\tilde{x}_{\mathrm{R}}(\tau), \tilde{u}(\cdot)) \le J_\Delta^*\big(x_{\mathrm{R}}^*(t_i), u_\Delta^*(\cdot;t_i)\big) - \int_{t_i}^{\tau} F_{\mathrm{R}}\big(x_\Delta^*(t;t_i), u_\Delta^*(t;t_i)\big)\,\mathrm{d}t\,.
$$

Since $x_{\mathrm{R}}^*(\tau)$, $u_\Delta^*(\cdot;\tau)$ is the optimal solution,

$$
\begin{aligned}
J_\Delta^*\big(x_{\mathrm{R}}^*(\tau), u_\Delta^*(\cdot;\tau)\big) &\le J_\Delta(\tilde{x}_{\mathrm{R}}(\tau), \tilde{u}(\cdot)) \\
&\le J_\Delta^*\big(x_{\mathrm{R}}^*(t_i), u_\Delta^*(\cdot;t_i)\big) - \int_{t_i}^{\tau} F_{\mathrm{R}}\big(x_\Delta^*(t;t_i), u_\Delta^*(t;t_i)\big)\,\mathrm{d}t\,.
\end{aligned}
$$

Since $x_\Delta(t) = x_\Delta^*(t;t_i)$ and $u_\Delta(t) = u_\Delta^*(t;t_i)$ for $t_i \le t < t_{i+1}$, we have shown (5.30).
By induction and nonnegativity of $J_\Delta^*\big(x_{\mathrm{R}}^*(t_i), u_\Delta^*(\cdot;t_i)\big)$ for $t_i \to \infty$ follows

$$
\int_{t_0}^{\infty} F_{\mathrm{R}}\big(x_\Delta(t), u_\Delta(t)\big)\,\mathrm{d}t \le J_\Delta^*\big(x_{\mathrm{R}}^*(t_0), u_\Delta^*(\cdot;t_0)\big)\,. \qquad (5.35)
$$

Combining (5.35) and F1) results in F2).

F3) Consider the null decision variables $x_{\mathrm{R}}^*(t_0) = 0$, $u_\Delta^*(\cdot;t_0) \equiv 0$. Associated with this choice, the error bound satisfies $\dot{\Delta}^*(t;t_0) = -\beta\Delta^*(t;t_0)$ due to (5.12e). This results in $\Delta^*(t;t_0) = \Delta^*(t_0;t_0)\,\mathrm{e}^{-\beta(t-t_0)}$ for all $t \in [t_0, t_0+T]$. Furthermore, consider an arbitrary initial condition $x(t_0)$ with $\|x(t_0)\| \le \gamma_\Delta/\alpha$. Then, $\Delta^*(t_0,t_0) \le \alpha\|x(t_0) - Vx_\Delta^*(t_0;t_0)\| = \alpha\|x(t_0)\| \le \gamma_\Delta$. Consequently, since Assumption 5.17 guarantees $\beta > 0$, the error bound satisfies $\Delta^*(t;t_0) \le$

$\gamma_\Delta \, \mathrm{e}^{-\beta(t-t_0)} \leq \gamma_\Delta$ for all $t \in [t_0, t_0+T]$. Besides, the null decision variables result in $x^*_\Delta(t;t_0)^\mathsf{T} P^\Omega_\Delta x^*_\Delta(t;t_0) = 0$ for all $t \in [t_0, t_0+T]$. Since Ω_Δ is of the form (5.21), we have $\big(x^*_\Delta(t;t_0), \Delta^*(t;t_0)\big) \in \Omega_\Delta$ for all $t \in [t_0, t_0+T]$. Since the state and input constraints are satisfied in the terminal set, (5.12c) and (5.12d) are fulfilled, showing that this choice of decision variables is a feasible solution if $\|x(t_0)\| \leq \gamma_\Delta/\alpha$. Since $J_\Delta\big(0,0\big) = 0$, for all $\|x(t_0)\| \leq \gamma_\Delta/\alpha$ the optimal solution of Problem 5.8 satisfies $J^*_\Delta\big(x^*_\mathrm{R}(t_0), u^*_\Delta(\cdot;t_0)\big) = 0$. Thus, from F2) follows $\|u_\Delta(\cdot)\|^2_{\mathcal{L}_2} = 0$. Hence, F3) is established.

i) From F3) follows that the origin is an equilibrium of the preprocessed model controlled by Δ-MPC. From Assumption 5.17 we know that the preprocessed model is asymptotically stable. Thus, the equilibrium $x = 0$ for the closed loop of the preprocessed system with the model predictive controller $u_\Delta(\cdot)$ given by Algorithm 5.9 is (locally) asymptotically stable.

ii) For every feasible solution of Problem 5.8, we observe that (5.12e), (5.12f) together with Theorem 4.4 guarantees $x(t) \in \mathcal{X}\big(x^*_\Delta(t;t_i), \Delta^*(t;t_i)\big)$ for all $t \in [t_i, t_i+T]$. Thus, according to Proposition 5.7, inequality (5.5) is guaranteed by (5.12c). Consequently, the state and input constraints (5.5) for the preprocessed model are satisfied when applying $u_\Delta(\cdot)$ to the preprocessed model.

iii) Let $\mathcal{F}^\Delta_\Delta \subseteq \mathbb{R}^n$ denote the set of all initial states $x(0)$ for which Problem 5.8 is feasible. Let $x(0) \in \mathcal{F}^\Delta_\Delta$. Then, the value of the cost functional for the optimal solution at time $t_0 = 0$ is finite. Hence, with (5.29) follows that $\|u_\Delta(\cdot)\|_{\mathcal{L}_2}$ is finite. From Assumption 5.17 we know that A is Hurwitz. Hence, the preprocessed model is finite gain $\mathcal{L}_2$ stable [Khalil, 1996, Theorem 6.4]. Thus, from the finite $\mathcal{L}_2$-norm of the input follows that $\|x(\cdot)\|_{\mathcal{L}_2}$ is finite. Furthermore, from Assumption 5.5 we conclude that both $x(t)$ and $u_\Delta(t)$ are bounded. Thus, $\frac{d}{dt}\|x(t)\|^2 = 2x^\mathsf{T}(t)\dot{x}(t) = 2x^\mathsf{T}(t)\big(Ax(t) + Bu_\Delta(t)\big)$ is bounded. Hence, $\|x(t)\|^2$ is uniformly continuous in t. Finally, by means of Barbalat's lemma [Khalil, 1996, Lemma 4.2] $\lim_{t\to\infty} \|x(t)\|^2 = 0$. Consequently, $\lim_{t\to\infty} x(t) = 0$. Thus, $\mathcal{F}^\Delta_\Delta$ is a subset of the region of attraction. If $x(0) \notin \mathcal{F}^\Delta_\Delta$, then $u_\Delta(t)$ is undefined and $x(0)$ is not in the region of attraction. Altogether, $\mathcal{F}^\Delta_\Delta$ and the region of attraction coincide. □

Theorem 5.22 states asymptotic stability of the preprocessed model controlled by Δ-MPC. Since the problem statement for the preprocessed model is derived from the problem statement for the plant by taking the preprocessing into account, from Theorem 5.22 follows that the origin of the closed loop of the plant with the prestabilizing controller K_P and Δ-MPC is asymptotically stable while satisfying the state and input constraints (5.2). Furthermore, the region of attraction of the plant in closed loop with the prestabilizing controller K_P, the state transformation T_P, and Δ-MPC is given by $\mathcal{F}_\Delta$ in which $\mathcal{F}_\Delta \subseteq \mathbb{R}^n$ denotes the set of all initial states $x_\mathrm{P}(t_0)$ such that Problem 5.8 is feasible for $x(t_0) = T_\mathrm{P} x_\mathrm{P}(t_0)$.

According to Theorem 5.22, the prediction horizon T should be chosen large enough such that Problem 5.8 is feasible at time $t = 0$. In particular, since the error bounding system is not controllable, a lower bound for the prediction horizon in dependence of the initial condition $x_0 = T_P x_{P,0}$, the decay rate of the error bound β, and the largest value of the error bound within the terminal constraint γ_Δ can be derived. The smallest possible initial error bound for any $x_{R,0}$ is $\|(I_n - \check{U}\check{U}^\mathsf{T})x_0\|$ in which $\check{U}\check{U}^\mathsf{T}$ is the orthogonal projection onto the span of the columns of V introduced in Proposition 5.10 on page 85. Thus, from the dynamics of the error bounding system follows that the predicted error bound at the prediction horizon is bounded from below $\Delta(T) \geq \mathrm{e}^{-\beta T} \|(I_n - \check{U}\check{U}^\mathsf{T})x_0\|$. Thus, Problem 5.8 is infeasible if $\mathrm{e}^{-\beta T} \|(I_n - \check{U}\check{U}^\mathsf{T})x_0\| \geq \gamma_\Delta$. Consequently, a necessary condition for feasibility of Problem 5.8 is

$$T \geq \frac{1}{\beta} \ln \left(\frac{\|(I_n - \check{U}\check{U}^\mathsf{T})x_0\|}{\gamma_\Delta} \right).$$

5.7 Relation to Existing Approaches

In this section, we compare the derived Δ-MPC scheme with existing MPC approaches using reduced models with robustness against the model reduction error on a theoretical basis.

MPC approaches using reduced models with robustness against the model reduction error are summarized in Table 1.1 on page 11. The Δ-MPC approach is only restricted to model reduction by projection. Hence, we allow for a large class of model reduction procedures, which encompasses many well-known model reduction methods for LTI systems as discussed in Section 2.1.1.

In contrast, the model reduction method is limited to modal truncation in [Dubljevic et al., 2006] and balanced truncation in [Narciso and Pistikopoulos, 2008]. In contrast to [Narciso and Pistikopoulos, 2008], we establish asymptotic stability of the origin. Furthermore, we prove under mild assumptions that the Δ-MPC scheme is recursively feasible. Hence, we do not have to assume recursive feasibility as in [Dubljevic et al., 2006].

In [Bäthge et al., 2016; Hovland et al., 2008b; Kögel and Findeisen, 2015; Lorenzetti et al., 2019; Sopasakis et al., 2013], any model reduction method is allowed, which is a minor advantage over the Δ-MPC approach. But, in contrast to [Bäthge et al., 2016; Hovland et al., 2008b], we use the a-posteriori bound for the model reduction error to establish satisfaction of hard state constraints without approximation of the model reduction error by an additive bounded uncertainty as in [Bäthge et al., 2016]. The main novelty of the derived Δ-MPC approach over the tube-based robust MPC approaches in [Kögel and Findeisen, 2015; Lorenzetti and Pavone, 2019; Lorenzetti et al., 2019; Sopasakis et al., 2013] is the utilized error bound: The a-posteriori error bound takes the actual input and state of the reduced model into account. In contrast, in [Kögel and Findeisen, 2015; Lorenzetti and Pavone, 2019; Lorenzetti et al., 2019; Sopasakis et al., 2013] sets are used that contain the error for all inputs and states in a bounded set. For large input and state constraints this results in a significant conservatism for the error bound. For example, the error bound of Δ-MPC is significantly less conservative than the error bound that takes the worst-case input and state into account for the tubular reactor as illustrated in Section 4.5.3. Furthermore, the a-posteriori error bound is given by an asymptotically stable dynamical system. As a consequence, the Δ-MPC guarantees asymptotic stability of the origin in contrast to a (possibly large) set around the origin in [Kögel and Findeisen, 2015; Lorenzetti et al., 2019; Sopasakis et al., 2013]. Asymptotic stability of the origin is also achieved in [Hovland et al., 2008b], but only for systems without state constraints. However, the a-posteriori error bound is a scalar bound on the norm of the error, which also introduces conservatism. Thus, for some applications a lower conservatism is possible by bounding the error with polytopic sets as in [Kögel and Findeisen, 2015; Lorenzetti et al., 2019; Sopasakis et al., 2013] or a bound for each face of the input and state constraints as in [Lorenzetti and Pavone, 2019].

Due to the dynamical error bound the Δ-MPC scheme is computationally more demanding than an MPC scheme using a constant error bound. The dynamical error bound increases the state dimension by one and, moreover, introduces a nonlinearity. In the following section, we investigate how large the computational burden of the dynamical error bound is by means of the tubular reactor. How the nonlinearity of the error bound affects the optimization problem is investigated in Chapter 6.

5.8 Example: Tubular Reactor

In this section, we continue with the tubular reactor to compare the three MPC schemes considered in this thesis as depicted in Table 5.1. The comparison with the $\mathcal{P}$-MPC scheme allows to assess the conservatism with respect to the performance of the proposed MPC scheme. For the tubular reactor, the order of the reduced model is chosen such that Problem 5.8 is feasible for all $\Delta T_{\text{in}} \in [-25\,\text{K}, 25\,\text{K}]$. Hence, the conservatism with respect to the region of attraction is not assessed. The comparison with the $\mathcal{R}$-MPC scheme shows the increase in the computational demand by adding the error bound and further adaptions in order to provide the guarantees of the Δ-MPC scheme.

First, in Section 5.8.1 we present the design of the model predictive controllers. Then, we assess the predicted and closed-loop behavior in Section 5.8.2, the computational efficiency in Section 5.8.3, and the conservatism of the Δ-MPC scheme compared to $\mathcal{P}$-MPC in Section 5.8.4.

In addition to the tubular reactor, also the academic Example 5.4 is used to compare the three MPC schemes. To assess the influence of a varying model reduction error, a parameter is added to the two-dimensional system. In the interest of brevity, the application to the academic example is presented in Appendix C and the main outcome is summarized in the following. The Δ-MPC scheme shows a good performance while guaranteeing asymptotic stability and constraint satisfaction, even when the closed loop with the $\mathcal{R}$-MPC scheme diverges. Furthermore, for this example, a reasonable region of attraction for Δ-MPC compared to $\mathcal{P}$-MPC is achieved by a suitable choice of the prestabilizing feedback.

The implementation is based on MATLAB R2012b and the MATLAB routine `nmpc` [Grüne and Pannek, 2017], which uses the MATLAB function `fmincon` for optimization. The ODEs are solved using the sundialsTB [Hindmarsh et al., 2005] with forward sensitivity analysis. The forward sensitivity analysis is used to provide the gradients of the cost functional and the constraints to `fmincon`. As optimization algorithm "active-set" within `fmincon` is chosen, since this results for the tubular reactor in the lowest computation time for all MPC schemes. The initial point for `fmincon` is based on the optimal solution of the previous sampling instant.

5.8.1 Design of the Model Predictive Controllers

The plant model and the control problem for the tubular reactor has been introduced in Section 2.4. For Δ-MPC, the reduced model of order 40 and error bound derived in Section 4.5 is used. Hence, the LQR and a state transformation is used in the preprocessing of the plant model. This results in $\alpha = 1$ and $\beta = 0.3$. The Δ-MPC scheme is based on the equivalent cost functional

$$\check{J}^{\mathrm{inf}}(x_0, u) := \int_0^\infty \check{F}\big(x(t), u(t)\big)\,\mathrm{d}t\,,$$

in which $\check{F}\big(x(t), u(t)\big) = u^\mathsf{T}(t) R u(t)$ is the equivalent stage cost introduced in Section 5.5. The prestabilization with the LQR and the equivalent cost functional are also used for $\mathcal{P}$-MPC and $\mathcal{R}$-MPC, since it decreases the computation time of $\mathcal{P}$-MPC. The reduced model for $\mathcal{R}$-MPC is computed in the same way as the one for Δ-MPC from the prestabilized plant model, i.e., the preprocessed model without state transformation.

All three models used for the prediction in the model predictive controllers are asymptotically stable. Hence, we use terminal controllers which are identical to zero. The terminal cost has to be chosen such that it upper bounds the integrated stage cost within the terminal region, see (2.11) and (5.22). Since the stage cost does not penalize the state, the integrated stage cost is zero in the terminal region. Hence, we use no terminal cost in all MPC schemes.

Polytopic positively invariant sets are difficult or impossible to compute for the dimension of the plant and the reduced models. Hence, ellipsoids centered at the setpoint are used as terminal sets for the state for all three MPC schemes. For Δ-MPC, the terminal set is of the form (5.21). For Δ-MPC and $\mathcal{R}$-MPC the volume of the ellipsoidal terminal set for the reduced state is maximized. For $\mathcal{P}$-MPC, the maximal volume ellipsoid could not be computed due to the large dimension of the system. Since the stage cost does not penalize any state, the unique positive semidefinite solution of the ARE is zero and cannot be used to define the shape of the terminal set. As an alternative, an ellipsoid based on the state transformation matrix T_{P} is used.

The three MPC schemes are implemented with a sampling time δ of 0.2 s and a prediction horizon $T := 80\,\delta = 16\,\mathrm{s}$ as in [Agudelo et al., 2007a,b]. The inputs are piecewise constant within each sampling time.

5.8.2 Time Response

Trajectories predicted at time $t = 0\,\mathrm{s}$ for the three MPC schemes for $T_{\mathrm{in}} = 365\,\mathrm{K}$ are depicted in Figure 5.5. The constraint for the maximal temperature inside the reactor is active for all MPC schemes at $t = 4.8\,\mathrm{s}$. To guarantee a maximal temperature of 400 K inside the reactor despite the model reduction error, Δ-MPC uses a lower predicted temperature for the first jacket. The predicted trajectories of $\mathcal{P}$-MPC and $\mathcal{R}$-MPC are almost identical.

Trajectories of the plant in closed loop with each of the three MPC schemes for the same inlet fluid temperature T_{in} of 365 K are shown in Figure 5.6. The predicted and closed-loop behavior is almost identical for $\mathcal{P}$-MPC, since there is no mismatch between the plant and the prediction model of $\mathcal{P}$-MPC and the relatively long prediction horizon. In contrast, the Δ-MPC scheme uses that the error bound gets smaller when a new measurement is taken into account. Thus, the first jacket temperature of the closed-loop system is larger than predicted. This results in a maximal temperature significantly closer to the constraints. Furthermore, while the predicted output concentration of $\mathcal{P}$-MPC and Δ-MPC differ, the closed-loop trajectories are almost indistinguishable.

The $\mathcal{R}$-MPC scheme results in a maximal temperature inside the reactor larger than 400 K. Hence, the constraints are violated and no feasible solution of the optimization problem can be found at $t = 5\,\mathrm{s}$. Thus, $\mathcal{R}$-MPC is not recursively feasible. This underpins that the guarantees provided by Δ-MPC are important.

5.8.3 Computational Complexity

The computation time required by the three MPC schemes is shown in Figure 5.7. Only the first optimization after the step of T_{in} at $t = 0$ is shown since the warm start results in a smaller computation time afterwards. The steps in the computation time correspond to the number of iterations. For $\mathcal{P}$-MPC and $\mathcal{R}$-MPC up to two iterations are required. For Δ-MPC up to three iterations are required for the optimization. Reasons for the larger number of iterations of Δ-MPC are the tightening of the constraints and the nonlinearity of the error bound.

The maximal computation time increases for $\mathcal{P}$-MPC by a factor of 6.7 and decreases for $\mathcal{R}$-MPC by a factor of 2.3 in comparison to Δ-MPC. Thus, the Δ-MPC scheme shows a sensible compromise between increased computational efficiency and guarantees for the closed loop.

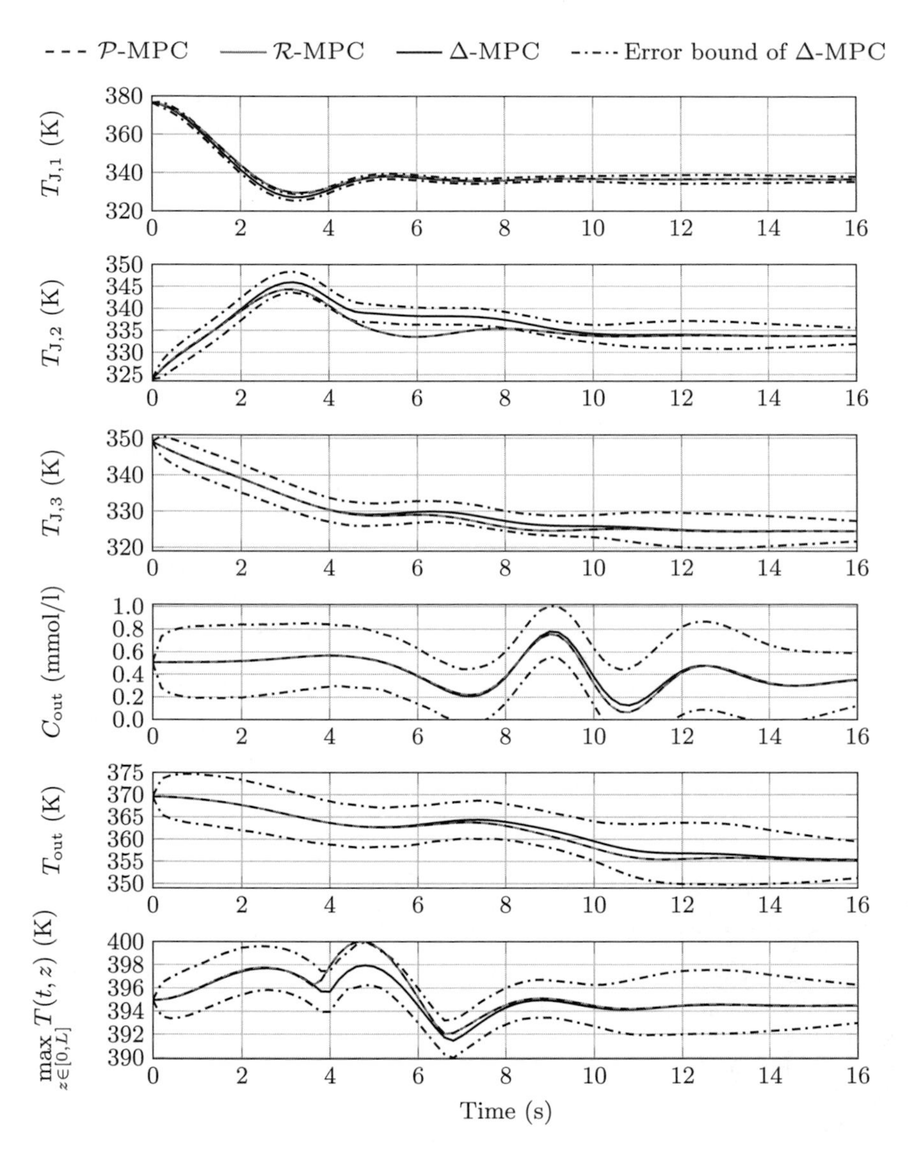

Figure 5.5: Trajectories predicted at time $t = 0\,\mathrm{s}$ for MPC using the plant model ($\mathcal{P}$-MPC), MPC using the reduced model ($\mathcal{R}$-MPC), and MPC using the reduced model and error bound (Δ-MPC) for $T_{\mathrm{in}} = 365\,\mathrm{K}$.

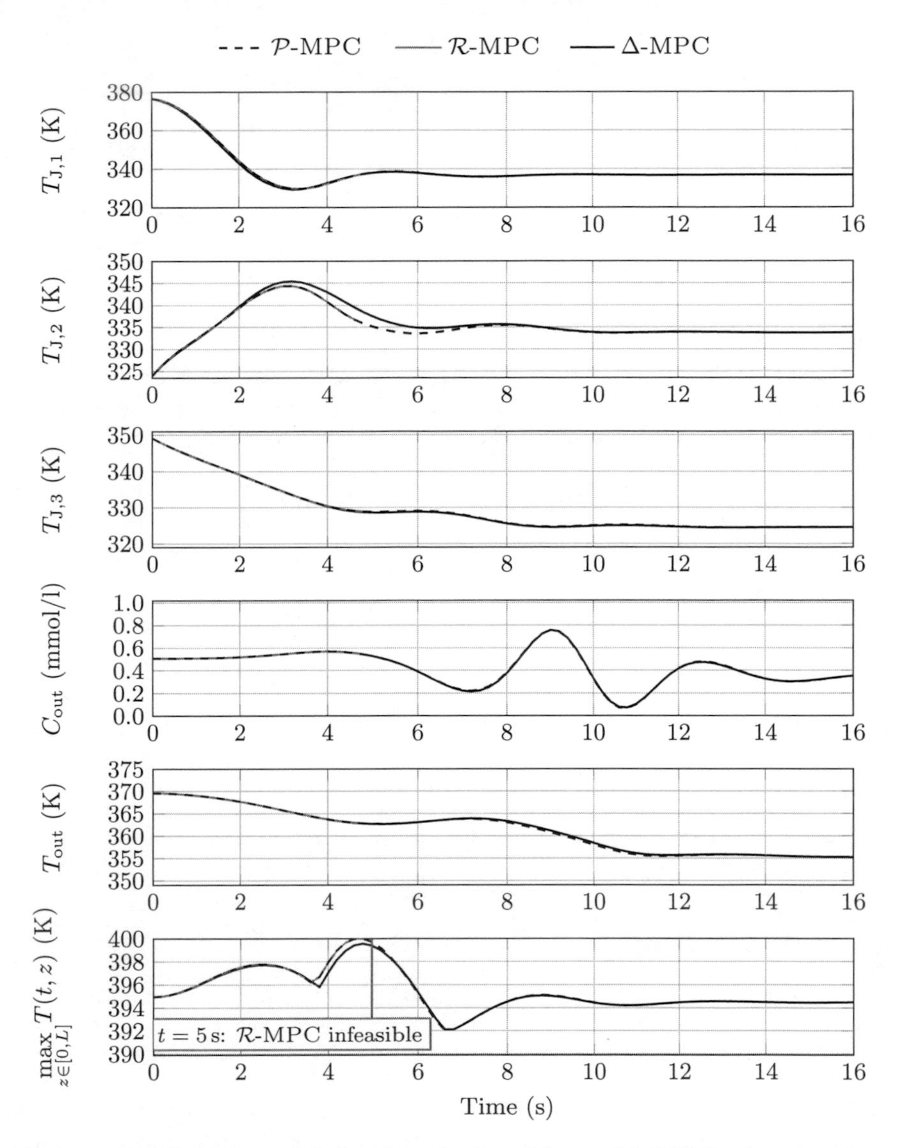

Figure 5.6: Trajectories of the plant in closed loop with MPC using the plant model ($\mathcal{P}$-MPC), MPC using the reduced model ($\mathcal{R}$-MPC), and MPC using the reduced model and error bound (Δ-MPC) for $T_{\text{in}} = 365\,\text{K}$.

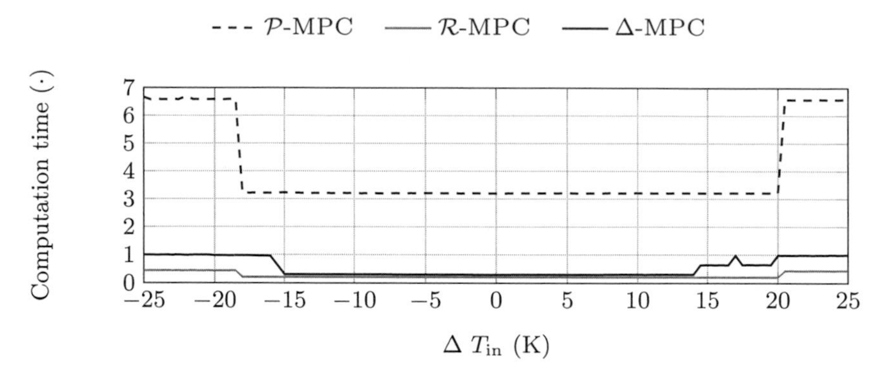

Figure 5.7: Computation time of $\mathcal{P}$-MPC, $\mathcal{R}$-MPC, and Δ-MPC for the optimization at $t = 0\,\text{s}$. The computation time is normalized by the maximal computation time of Δ-MPC, i.e., 24.7 s to ease comparison.

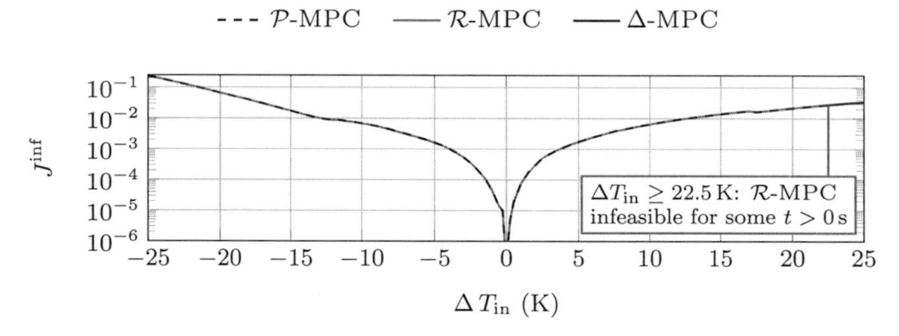

Figure 5.8: Cost functional values for the plant in closed loop with $\mathcal{P}$-MPC, $\mathcal{R}$-MPC, and Δ-MPC.

5.8.4 Performance

Finally, the performance of $\mathcal{P}$-MPC, $\mathcal{R}$-MPC, and Δ-MPC is compared. Figure 5.8 shows the value of the infinite horizon cost for the plant in closed loop with each of the three model predictive controllers when the inlet fluid temperature changes. There are only minor differences in the cost functional value of the three MPC schemes. Only for $\Delta T_{\text{in}} \geq 22.5\,\text{K}$ the $\mathcal{R}$-MPC scheme is feasible at $t = 0\,\text{s}$ but infeasible for some $t > 0\,\text{s}$ underpinning the lack of guarantees if only a reduced model is used for the prediction.

The performance degradation of using Δ-MPC instead of $\mathcal{P}$-MPC is depicted in Figure 5.9. The discontinuity at $\Delta T_{\text{in}} = 17.2\,\text{K}$ originates from the dependence of the steady state on T_{in} by means of Problem 2.15.

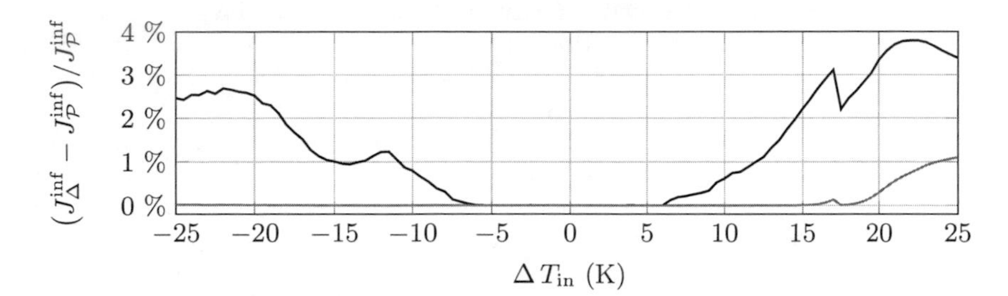

Figure 5.9: Relative performance degradation for the plant in closed loop with Δ-MPC compared to $\mathcal{P}$-MPC.

If Δ-MPC is based on the equivalent cost functional $\check{J}^{\mathrm{inf}}(x_0, u)$, then the performance of Δ-MPC is maximally 1.2% worse than $\mathcal{P}$-MPC, which is surprisingly low compared to the increase in computational efficiency. Furthermore, for an inlet fluid temperature with $\Delta T_{\mathrm{in}} \in [-16\,\mathrm{K}, 14.5\,\mathrm{K}]$ the performance of Δ-MPC and $\mathcal{P}$-MPC are identical up to the tolerance of the cost functional value used for the optimization. This demonstrates that the statement in Section 5.5 about the local optimality of Δ-MPC when using the equivalent cost functional has a practical relevance.

When using the preprocessed stage cost (5.6) instead of the equivalent stage cost, then the performance gets worse as depicted in Figure 5.9. One reason for this large difference is the optimization of the initial condition of the reduced model, which results in a bad approximation of the stage cost for the preprocessed stage cost. This comparison illustrates the practical relevance of Lemma 5.11.

In summary, this simulation study showed that Δ-MPC increases the computational efficiency by a factor of 6.7 compared to an MPC scheme based on the plant model, while at the same time resulting in a small conservatism with respect to performance and achieving the desired region of attraction. The comparison of the performance for the equivalent with the preprocessed cost functional demonstrated the practical relevance of Lemma 5.11.

5.9 Summary

In Section 5.1, we have demonstrated by an example that the $\mathcal{R}$-MPC scheme leads to an unstable closed loop with the plant although the $\mathcal{P}$-MPC scheme stabilizes the plant. Hence, simply using a reduced model for the prediction results in an MPC scheme that lacks guarantees for the closed loop with the plant. Hence, we presented a novel MPC scheme using a reduced model for the prediction that utilizes the results of Chapter 4, i.e., the bound for the model reduction error together with the preprocessing of the plant.

The main result of this chapter is stated in Theorem 5.22 and ensures asymptotic stability and constraint satisfaction for the plant in closed loop with the proposed MPC scheme under mild assumptions. Moreover, we derived constructive design conditions for the terminal set and terminal cost satisfying the assumptions of Theorem 5.22. For Δ-MPC recursive feasibility is established despite the model reduction error by using the initial state of the reduced model as decision variable and appropriate initialization of the error bounding system. Computational efficiency is gained by the computationally tractable formulation for the constraint satisfaction of the preprocessed model stated in Proposition 5.7.

When using the LQR for prestabilization of the plant, we showed that an equivalent cost functional eliminates the influence of the model reduction error on the infinite and finite horizon cost functional for a common choice of the terminal costs of $\mathcal{P}$-MPC and Δ-MPC. As a consequence, the Δ-MPC scheme implicitly minimizes the quasi-infinite horizon cost functional of the $\mathcal{P}$-MPC scheme. Furthermore, the proposed MPC scheme guarantees an upper bound for the cost functional value of the closed loop with the plant under mild assumptions in the considered case. In order to show asymptotic stability also for the equivalent cost functional, we appropriately relaxed the assumption on the cost functional for the reduced model.

The simulation study of controlling the tubular reactor showed that the Δ-MPC scheme significantly reduces the maximal computation time compared to the MPC scheme based on the plant model and results in minor differences in the performance with respect to the cost functional value. Moreover, by choosing the LQR for prestabilization, Δ-MPC results in an optimal solution for a large region around the equilibrium. This also emphasizes the practical relevance of the equivalent cost functional that is independent of the model reduction error. Altogether, the example demonstrated that Δ-MPC achieves a reasonable trade-off between computational efficiency and conservatism with respect to performance and region of attraction.

Chapter 6

Model Predictive Control Using Reduced Models with Guaranteed Properties for Discrete-Time Systems

In the previous chapter, a novel MPC scheme using a reduced model for the prediction has been presented for continuous-time systems. To apply the MPC scheme to the tubular reactor, in Chapter 5 only the input trajectory has been discretized in time. In industrial applications, also the plant model is typically discretized in time — especially when state constraints hinder the approach relying on the optimality conditions of the optimal control problem also known as the indirect method [Dunn, 1996; Graichen and Käpernick, 2012]. The time discretization of the plant model and interpretation of the difference equations as equality constraints results in a finite-dimensional *static* optimization problem. This typically reduces the computation time and eases the implementation since available solvers for static optimization problems can be used. In contrast, in Chapter 5 a variable-step ODE solver and the gradients for the cost function and the constraints have been used. Furthermore, considering discrete-time systems allows to compare the class of the optimization problems corresponding to the proposed MPC scheme and the two MPC schemes introduced in Section 2.3.

The focus of the previous chapter has been the theoretical properties of the proposed MPC scheme. The main goal of this chapter is an efficient formulation of the optimization problem for discrete-time systems.

This chapter starts with the problem statement in Section 6.1. Afterwards, the MPC scheme using the reduced model and error bound is introduced in Section 6.2. In both sections, the discrete-time equivalent of the continuous-time version in Chapter 5 is briefly stated. The main results of this chapter are presented in Section 6.3: At the beginning the optimization problem is reformulated as an equivalent convex optimization problem. Afterwards, the convex optimization problem is transformed such that it is independent of the dimension of the preprocessed model. Finally, the optimization problem is stated as a second-order cone program (SOCP). Thus, dedicated and more efficient solvers are available to compute the solution. In Section 6.4, the computational demand of the considered MPC schemes is assessed by means of the tubular reactor.

6.1 Problem Statement

In this chapter, we consider discrete-time LTI models described by

$$\Sigma_{\mathrm{PP}}:\begin{cases} x(i+1) = Ax(i) + Bu(i)\,, \\ x(0) = x_0\,, \end{cases} \tag{6.1}$$

in which $i \in \mathbb{N}_0$ denotes the sample number corresponding to the time $t = i\delta$, δ is the sampling time, $x(i) \in \mathbb{R}^n$ is the state vector, $u(i) \in \mathbb{R}^{n_{\mathrm{U}}}$ is the input vector, and $x_0 \in \mathbb{R}^n$ is the initial condition. The model order n is assumed to be large, i.e., $n \gg 1$. For the sake of brevity, we assume that the model (6.1) results from the preprocessing of a plant model similar to the procedure described for continuous-time systems in Section 5.2. Hence, we assume that the model (6.1) is asymptotically stable and satisfies the following assumption.

Assumption 6.1. *The norm of the i-th power of the state matrix A of the preprocessed model* (6.1) *is bounded by*

$$\left\|A^i\right\| \leq \alpha\beta^i \quad \textit{for all } i \in \mathbb{N}_0\,,$$

with $\alpha \geq 1$ *and* $0 < \beta < 1$.

The assumption is not restrictive since for all asymptotically stable preprocessed models the spectral radius ρ of the state matrix A is strictly smaller than 1. In this case, there exists an $\alpha \geq 1$ and $\rho(A) < \beta < 1$ satisfying Assumption 6.1.

The states and inputs of the preprocessed model are constrained to a polytopic set by

$$c_k^{\mathsf{T}}\begin{bmatrix} x(i) \\ u(i) \end{bmatrix} \leq d_k\,, \qquad k = 1, \ldots, n_{\mathrm{C}} \tag{6.2}$$

in which $c_k \in \mathbb{R}^{n+n_{\mathrm{U}}}$ and $d_k \in \mathbb{R}$ for all $i \in \mathbb{N}_0$.

Given the initial state x_0, we want to steer the system to the origin close to optimality with respect to the infinite horizon cost function

$$J^{\inf}(x_0, u) := \sum_{i=0}^{\infty} F\big(x(i), u(i)\big)\,,$$

subject to the system dynamics (6.1) as well as the state and input constraints (6.2). A common cost function is the quadratic stage cost that is defined by

$$F\big(x(i), u(i)\big) := x^{\mathsf{T}}(i)Qx(i) + 2x^{\mathsf{T}}(i)Su(i) + u^{\mathsf{T}}(i)Ru(i)\,, \tag{6.3}$$

in which Q, S, and R are weighting matrices of appropriate dimension. We assume that the full state of the preprocessed model can be measured.

We seek an approximate solution of the control problem via MPC using a reduced model and an a-posteriori error bound similar to the bound given in Theorem 4.4. We allow for any model reduction method based on a projection of the preprocessed model, which is a very common framework including most of the methods described in Antoulas [2005b]. Hence, we assume that the matrices V and W are given and result in the reduced model

$$\Sigma_{\mathrm{R}}: \begin{cases} x_{\mathrm{R}}(i+1) = A_{\mathrm{R}} x_{\mathrm{R}}(i) + B_{\mathrm{R}} u(i), \\ \quad x_{\mathrm{R}}(0) = x_{\mathrm{R},0}, \end{cases} \tag{6.4}$$

in which $A_{\mathrm{R}} = W^{\mathsf{T}} A V$, $B_{\mathrm{R}} = W^{\mathsf{T}} B$. The stage cost is

$$F_{\mathrm{R}}\big(x_{\mathrm{R}}(i), u(i)\big) := x_{\mathrm{R}}^{\mathsf{T}}(i) Q_{\mathrm{R}} x_{\mathrm{R}}(i) + 2 x_{\mathrm{R}}^{\mathsf{T}}(i) S_{\mathrm{R}} u(i) + u^{\mathsf{T}}(i) R_{\mathrm{R}} u(i),$$

in which $Q_{\mathrm{R}} = V^{\mathsf{T}} Q V$, $S_{\mathrm{R}} = V^{\mathsf{T}} S$, and $R_{\mathrm{R}} = R$. In the remainder of this chapter, we restrict the parameters of the stage cost by the following assumption.

Assumption 6.2. *The matrix R_{R} is symmetric and positive definite. Furthermore, the matrices Q_{R} and S_{R} satisfy at least one of the following statements:*

i) The matrix $Q_{\mathrm{R}} - S_{\mathrm{R}} R_{\mathrm{R}}^{-1} S_{\mathrm{R}}^{\mathsf{T}}$ is symmetric and positive definite.

ii) The matrix Q_{R} is symmetric and positive semidefinite and $S_{\mathrm{R}} = 0_{n_{\mathrm{R}} \times n_{\mathrm{U}}}$.

Similar to Section 5.5, the second case allows to use the equivalent stage cost for the preprocessed model

$$\check{F}\big(x(i), u(i)\big) = u^{\mathsf{T}}(i)\Big(R + B^{\mathsf{T}} P B\Big) u(i),$$

in which P is the solution of the discrete-time Lyapunov equation $P - A^{\mathsf{T}} P A - Q = 0_{n \times n}$, if the stabilizing LQR is used in the preprocessing of the plant [Chisci et al., 2001; Kouvaritakis et al., 2002].

6.2 MPC Scheme Using the Reduced Model and Error Bound

For the setup introduced in Section 6.1, we have the following bound for the model reduction error.

Theorem 6.3 (A-posteriori error bound for discrete-time systems)**.** *Consider the preprocessed model* (6.1)*, the reduced model* (6.4) *defined by V, W and $\alpha \geq 1$ and $0 < \beta < 1$ satisfying Assumption 6.1. Then, the model reduction error is bounded by*

$$\|x(i) - V x_{\mathrm{R}}(i)\| \leq \Delta(i) \quad \textit{for all } i \in \mathbb{N}_0,$$

in which $\Delta(i)$ is defined by the error bounding system

$$\Sigma_{\Delta}: \begin{cases} \Delta(i+1) = \beta \Delta(i) + \alpha \|r(i)\|, \\ \quad \Delta(0) = \alpha \|x(0) - V x_{\mathrm{R}}(0)\|, \end{cases} \tag{6.5}$$

with the residual $r(i) = \big(I_n - V W^{\mathsf{T}}\big)\big(A V x_{\mathrm{R}}(i) + B u(i)\big)$.

Proof. The proof is similar to the proof of Theorem 4.4 and [Haasdonk and Ohlberger, 2011, Proposition 1].

To derive the error bound, consider the error

$$e(i) = x(i) - Vx_{\mathrm{R}}(i)$$

between the state of the preprocessed model $x(i)$ and the estimate $Vx_{\mathrm{R}}(i)$. The error dynamics are given by

$$\begin{aligned} e(i+1) &= x(i+1) - Vx_{\mathrm{R}}(i+1) \\ &= Ax(i) + Bu(i) - VA_{\mathrm{R}}x_{\mathrm{R}}(i) - VB_{\mathrm{R}}u(i) \\ &= Ae(i) + \big(AV - VA_{\mathrm{R}}\big)x_{\mathrm{R}}(i) + \big(B - VB_{\mathrm{R}}\big)u(i) \\ &= Ae(i) + r(i)\,. \end{aligned}$$

The error dynamics have the solution

$$e(i) = A^i e(0) + \sum_{\tau=0}^{i-1} A^{i-\tau-1} r(\tau)\,.$$

By applying the norm and using the sub-multiplicative property we obtain

$$\|e(i)\| \leq \left\|A^i\right\| \|e(0)\| + \sum_{\tau=0}^{i-1} \left\|A^{i-\tau-1}\right\| \|r(\tau)\|\,.$$

Using Assumption 6.1, the error is bounded from above $\|e(i)\| \leq \Delta(i)$ by

$$\Delta(i) = \alpha\beta^i \|e(0)\| + \alpha \sum_{\tau=0}^{i-1} \beta^{i-\tau-1} \|r(\tau)\|\,.$$

To conclude the proof, we observe that the error bound $\Delta(i)$ is the solution of the error bounding system (6.5). □

With this error bound the state and input constraints (6.2) can be enforced, when using the reduced model (6.4) for the prediction. In the following, we state the Δ-MPC approach for discrete-time systems.

At every sampling instant $i \in \mathbb{N}_0$ the following finite horizon optimal control problem is solved.

Problem 6.4 (Optimization problem for MPC using the reduced model and error bounding system).

$$\underset{\bar{u},x_{\mathrm{R}}(i)\in\mathcal{I},\bar{x}_{\mathrm{R}},\bar{\Delta}}{\text{minimize}} \quad J_\Delta\big(x_{\mathrm{R}}(i),\bar{u}(\cdot;i)\big) := \sum_{\tau=i}^{i+T-1} F_{\mathrm{R}}\big(\bar{x}_{\mathrm{R}}(\tau;i),\bar{u}(\tau;i)\big) + E_\Delta\big(\bar{x}_{\mathrm{R}}(i+T;i)\big)$$

subject to

$$\bar{x}_{\mathrm{R}}(\tau+1;i) = A_{\mathrm{R}}\bar{x}_{\mathrm{R}}(\tau;i) + B_{\mathrm{R}}\bar{u}(\tau;i)\,, \tag{6.6a}$$

$$\bar{x}_{\mathrm{R}}(i;i) = x_{\mathrm{R}}(i)\,, \tag{6.6b}$$

$$c_k^{\mathsf{T}}\begin{bmatrix} V\bar{x}_{\mathrm{R}}(\tau;i) \\ \bar{u}(\tau;i)\end{bmatrix} \le d_k - \bar{\Delta}(\tau;i)\left\| c_k^{\mathsf{T}}\begin{bmatrix} I_n \\ 0_{n_{\mathrm{U}}\times n}\end{bmatrix}\right\|, \qquad k=1,\dots,n_{\mathrm{C}}\,, \tag{6.6c}$$

$$\big(\bar{x}_{\mathrm{R}}(i+T;i),\bar{\Delta}(i+T;i)\big) \in \Omega_\Delta\,, \tag{6.6d}$$

$$\bar{\Delta}(\tau+1;i) = \beta\bar{\Delta}(\tau;i) + \alpha\left\|\Big(I_n - VW^{\mathsf{T}}\Big)\big(AV\bar{x}_{\mathrm{R}}(\tau;i)+B\bar{u}(\tau;i)\big)\right\|, \tag{6.6e}$$

$$\bar{\Delta}(i;i) = \begin{cases} \Delta^*(i;i-1) & \text{if } x_{\mathrm{R}}(i) = x_\Delta^*(i;i-1) \text{ and} \\ & \Delta^*(i;i-1) < \alpha\|x(i)-Vx_{\mathrm{R}}(i)\|\,, \\ \alpha\|x(i)-Vx_{\mathrm{R}}(i)\| & \text{otherwise}\,, \end{cases} \tag{6.6f}$$

for all $\tau = i,\dots,i+T-1$,

The structure of the discrete-time Δ-MPC scheme is equivalent to the continuous-time version and visualized in Figure 5.3 on page 80.

The input trajectory and initial condition of the reduced model that solve Problem 6.4 is denoted by $u_\Delta^*(\tau;i)$ and $x_{\mathrm{R}}^*(i)$ with associated predicted state of the reduced model $x_\Delta^*(\tau;i)$ and predicted error bound $\Delta^*(\tau;i)$ for $\tau=i,\dots,i+T$.

The discrete-time Δ-MPC scheme is given by the following algorithm.

Algorithm 6.5 (Δ-MPC: MPC using the reduced model and error bounding system for discrete-time systems).

Require: *The preprocessed model* Σ_{PP} (6.1), *the matrices* V *and* W *defining the reduced model* Σ_{R} (6.4), *the constraints* (6.2), *the error bounding system* Σ_Δ (6.5) *with* $\alpha \ge 1$ *and* $0<\beta<1$, *the sampling time* $0<\delta\le T$, *the matrices defining the stage cost* Q_{R}, S_{R}, R, *the terminal cost* $E_\Delta(\cdot)$, *and the terminal set* Ω_Δ

$i \leftarrow 0$

loop

$x(i) \leftarrow$ *Measure the state of the preprocessed model at time* $t=i\delta$

$u_\Delta^*(\cdot;i)$, $x_\Delta^*(i+1;i)$, $\Delta^*(i+1;i) \leftarrow$ *Solve Problem 6.4*

$u(\cdot) = u_\Delta(\cdot) \leftarrow$ *Apply* $u_\Delta^*(i;i)$ *for* $t\in[i\delta,(i+1)\delta)$ *to* Σ_{PP}

$i \leftarrow i+1$

end loop

Satisfaction of the state and input constraints directly follows from the a-posteriori error bound of Theorem 6.3 and the reformulation of the constraints stated in Proposition 5.7. The guarantees stated in Lemma 5.20, Theorem 5.21, and Theorem 5.22 are valid equivalently for the discrete-time Δ-MPC scheme. The proofs are essentially identical to the proofs for continuous-time systems. The main difference is that for the invariance of the terminal set the parameters γ_{x} and γ_Δ have to be chosen such that

$$\alpha\sqrt{\gamma_{\mathrm{x}}}\left\|\left(I_n - VW^{\mathsf{T}}\right)\left(AV - BK_\Omega\right)\left(P_\Delta^\Omega\right)^{-1/2}\right\| \leq (1-\beta)\,\gamma_\Delta\,.$$

This results in

$$\Delta(i+1) - \Delta(i) \leq (\beta - 1)(\Delta(i) - \gamma_\Delta)\,.$$

Since $0 < \beta < 1$, the set $\{\Delta \in \mathbb{R} \mid 0 \leq \Delta \leq \gamma_\Delta\}$ is positive invariant. Furthermore, Barbalat's Lemma is not required to prove the region of attraction: From $\|x(\cdot)\|_{\ell_2}$ being finite it follows $\lim_{i\to\infty} x(i) = 0$.

The distinction of cases in (6.6f) can be circumvented by solving two optimization problems, one for each case in (6.6f), and using the solution with the lower cost function value. The optimization problem for the first case needs only to be solved if $\Delta^*(i; i-1) < \alpha\|x(i) - Vx_\Delta^*(i; i-1)\|$. If $\alpha = 1$, then the error bounding system guarantees that this cannot happen. Hence, only the optimization problem for the second case needs to be solved for $\alpha = 1$. The second case leads to the following optimization problem.

Problem 6.6 (Optimization problem for Δ-MPC with reinitialized error bound)**.**

$$\begin{aligned}
&\underset{\bar{u},x_{\mathrm{R}}(i)\in\mathcal{I},\bar{x}_{\mathrm{R}},\bar{\Delta}}{\text{minimize}} && J_\Delta\left(x_{\mathrm{R}}(i), \bar{u}(\cdot; i)\right) \\
&\text{subject to} && \text{(6.6a)–(6.6e) } \textit{and} \\
& && \bar{\Delta}(i; i) = \alpha\|x(i) - Vx_{\mathrm{R}}(i)\| \qquad (6.7a)\\
& && \text{for all } \tau = i, \ldots, i+T-1\,.
\end{aligned}$$

In the following, we consider only Problem 6.6, which has to be solved at every sampling instant. The results of the subsequent section are also valid for the optimization problem of the first case by replacing (6.7a) with the linear equality constraints $\bar{\Delta}(i; i) = \Delta^*(i; i-1)$ and $x_{\mathrm{R}}(i) = x_\Delta^*(i; i-1)$.

6.3 Equivalence of the Optimization Problem to a Second-Order Cone Problem

To apply the Δ-MPC scheme introduced in the previous section, the solution of Problem 6.6 has to be computed efficiently. Due to the nonaffine equality constraints (6.6e) and (6.7a), Problem 6.6 is not a convex optimization problem [Boyd and Vandenberghe, 2004]. In this section, we reformulate the optimization problem into an SOCP with low-dimensional constraints in order to facilitate the optimization.

In the remainder of this chapter, we require the following assumptions on the terminal set and terminal cost.

Assumption 6.7. *The terminal set Ω_Δ is of the form*

$$\Omega_\Delta := \left\{(x_\mathrm{R}, \Delta) \in \mathbb{R}^{n_\mathrm{R}} \times \mathbb{R} \;\middle|\; x_\mathrm{R}^\mathsf{T} P_\Delta^\Omega x_\mathrm{R} \leq \gamma_\mathrm{x},\, 0 \leq \Delta \leq \gamma_\Delta \right\}, \tag{6.8}$$

in which $P_\Delta^\Omega \succ 0$.

Assumption 6.8. *The terminal cost is given by $E_\Delta(x_\mathrm{R}) := x_\mathrm{R}^\mathsf{T} Q_\Delta^\Omega x_\mathrm{R}$ in which $Q_\Delta^\Omega \succeq 0$.*

Hence, the terminal constraint (6.6d) can be written in terms of a quadratic constraint and two linear inequalities

$$\bar{x}_\mathrm{R}^\mathsf{T}(i+T;i) P_\Delta^\Omega \bar{x}_\mathrm{R}(i+T;i) \leq \gamma_\mathrm{x}\,, \tag{6.9a}$$

$$0 \leq \bar{\Delta}(i+T;i) \leq \gamma_\Delta\,. \tag{6.9b}$$

Both, the cost function J_Δ and the inequality constraints (6.6c) and (6.9) are convex with respect to the optimization variables. To end up with affine equality constraints, we relax the optimization problem by replacing the equality constraints (6.6e) and (6.7a) with

$$\alpha \left\| \left(I_n - VW^\mathsf{T}\right)\left(AV\bar{x}_\mathrm{R}(\tau;i) + B\bar{u}(\tau;i)\right) \right\| \leq \bar{\Delta}(\tau+1;i) - \beta\bar{\Delta}(\tau;i)\,, \tag{6.10a}$$

$$\alpha \|x(i) - Vx_\mathrm{R}(i)\| \leq \bar{\Delta}(i;i)\,. \tag{6.10b}$$

If $\beta \geq 0$, then the relaxation of the constraints allows for increasing the error bound $\Delta(\cdot;i)$, which results in an additional tightening of the state and input constraints (6.6c).

The reformulated optimization problem is given by the following optimization problem.

Problem 6.9 (Convex optimization problem for Δ-MPC)**.**

$$\begin{aligned} \underset{\bar{u}, x_\mathrm{R}(i) \in \mathcal{I}, \bar{x}_\mathrm{R}, \bar{\Delta}}{\text{minimize}} \quad & J_\Delta\left(x_\mathrm{R}(i), \bar{u}(\cdot;i)\right) \\ \text{subject to} \quad & \text{(6.6a)–(6.6c), (6.9), } \textit{and } \text{(6.10)} \\ & \text{for all } \tau = i, \dots, i+T-1\,. \end{aligned}$$

The constraints (6.6e) and (6.7a) of Problem 6.6 have been relaxed to end up with Problem 6.9. How the two optimization problems are related, is stated below in Proposition 6.10. Beforehand, we define that two optimization problems are denoted equivalent, if from an optimal solution of one problem an optimal solution of the other problem can be directly computed, and the other way around [Boyd and Vandenberghe, 2004].

Proposition 6.10. *Consider the preprocessed model* (6.1)*, the reduced model* (6.4) *defined by V, W, and the error bounding system* (6.5)*. Suppose that Assumptions 6.1, 6.2, 6.7, and 6.8 are satisfied. Then, Problem 6.6 and Problem 6.9 are equivalent.*

Proof. Using that the terminal set Ω_Δ is of the form (6.8), the constraints (6.6d) and (6.9) are equivalent. Thus, the difference between the two optimization problems originates from relaxing the equality constraints (6.6e) and (6.7a) with the inequality constraints (6.10).

First, we show that a feasible solution of one optimization problem can be computed from the optimal solution of the other optimization problem. Let $u^*_\Delta(\cdot;i)$, $x^*_\mathrm{R}(i)$, $x^*_\Delta(\cdot;i)$, $\Delta^*(\cdot;i)$ be an optimal solution of Problem 6.9 with associated cost $J^*_\Delta(x(i))$. Consider $\Delta^\mathrm{e}(\cdot;i)$ satisfying the equality constraints (6.6e) and (6.7a) for $u^*_\Delta(\cdot;i)$, $x^*_\mathrm{R}(i)$, and $x^*_\Delta(\cdot;i)$. Since $\beta > 0$, with (6.10) we have $\Delta^\mathrm{e}(\tau;i) \leq \Delta^*(\tau;i)$ for all $\tau = i, \ldots, i+T$. Thus, the state and input constraints (6.6c) are satisfied for $u^*_\Delta(\cdot;i)$, $x^*_\Delta(\cdot;i)$ and $\Delta^\mathrm{e}(\cdot;i)$. Furthermore, since $\Delta^\mathrm{e}(\cdot;i) \geq 0$, the terminal constraint for the error bound (6.9b) is fulfilled for $\Delta^\mathrm{e}(\cdot;i)$. Thus, the modified solution $u^*_\Delta(\cdot;i)$, $x^*_\Delta(\cdot;i)$, $\Delta^\mathrm{e}(\cdot;i)$ is a feasible solution of Problem 6.6. Furthermore, an optimal solution of Problem 6.6 is also a feasible solution of Problem 6.9, since the constraints have been relaxed.

It remains to show that the computed feasible solution is also an optimal solution. The constraints in Problem 6.6 are tighter than the constraints in Problem 6.9. Furthermore, Assumption 6.2 and 6.8 guarantee that the cost function is convex. Hence, an optimal cost function value of Problem 6.6 is larger or equal to $J^*_\Delta(x(i))$. Hence, $x^*_\Delta(i)$, u^*_Δ, $x^*_\Delta(\cdot;i)$, and $\Delta^\mathrm{e}(\cdot;i)$ is an optimal solution of Problem 6.6.

Since $0 < \beta$, replacing the equality constraints (6.6e) and (6.7a) with the equality and inequality constraints (6.10) allows only for increasing the error bound $\Delta(\cdot;i)$. Increasing the value of the error bound results in an additional tightening of the state and input constraints (6.6c). Furthermore, the error bound is only coupled with the state and input by the state and input constraints (6.6c). Since, in addition, the cost function is independent of the error bound, the value of the cost function can only increase by replacing the equality constraints (6.6e) and (6.7a) with the equality and inequality constraints (6.10). Thus, the optimal solution of Problem 6.6 is also an optimal solution of Problem 6.9.

Altogether, the two optimization problems are equivalent. □

Problem 6.9 is a convex optimization problem with a quadratic objective function, affine equality and inequality constraints, one quadratic constraint, and $T+1$ second-order cone constraints. The quadratic constraint (6.9a) can be written as a second-order cone constraint by taking the square root. This results in (6.13d) below.

The second-order cone constraints (6.10a) and (6.10b) are of dimension $n+1$. In the following, we reformulate these constraints to end up with an optimization problem in which the dimensions of the constraints are independent of the dimension of the preprocessed model.

For the constraint (6.10a) the SVD is used to compute a full-rank factorization [Piziak and Odell, 1999] of $\tilde{R} := (I_n - VW^\mathsf{T})[AV \quad B] = \tilde{U}\tilde{\Sigma}\tilde{V}^\mathsf{T}$, in which $\tilde{U}^\mathsf{T}\tilde{U} = I_{\tilde{r}}$, $\tilde{V}^\mathsf{T}\tilde{V} = I_{\tilde{r}}$, $\tilde{\Sigma} = \operatorname{diag}(\sigma_1(\tilde{R}), \dots, \sigma_{\tilde{r}}(\tilde{R}))$, and $\tilde{r} = \operatorname{rank}(\tilde{R}) \le n_\mathrm{R} + n_\mathrm{U}$. Then, the constraint (6.10a) can be reformulated.

$$\begin{aligned}
&\alpha\left\|\left(I_n - VW^\mathsf{T}\right)[AV \quad B]\begin{bmatrix}\bar{x}_\mathrm{R}(\tau;i)\\ \bar{u}(\tau;i)\end{bmatrix}\right\| = \alpha\left\|\tilde{R}\begin{bmatrix}\bar{x}_\mathrm{R}(\tau;i)\\ \bar{u}(\tau;i)\end{bmatrix}\right\| \le \bar{\Delta}(\tau+1;i) - \beta\bar{\Delta}(\tau;i)\\
&\iff \alpha\left(\begin{bmatrix}\bar{x}_\mathrm{R}^\mathsf{T}(\tau;i) & \bar{u}^\mathsf{T}(\tau;i)\end{bmatrix}\tilde{V}\tilde{\Sigma}^2\tilde{V}^\mathsf{T}\begin{bmatrix}\bar{x}_\mathrm{R}(\tau;i)\\ \bar{u}(\tau;i)\end{bmatrix}\right)^{1/2} \le \bar{\Delta}(\tau+1;i) - \beta\bar{\Delta}(\tau;i)\\
&\iff \alpha\left\|\tilde{\Sigma}\tilde{V}^\mathsf{T}\begin{bmatrix}\bar{x}_\mathrm{R}(\tau;i)\\ \bar{u}(\tau;i)\end{bmatrix}\right\| \le \bar{\Delta}(\tau+1;i) - \beta\bar{\Delta}(\tau;i)\,.
\end{aligned} \tag{6.11}$$

For the constraint (6.10b) consider the SVD of the full-column rank matrix $V = \check{U}\check{\Sigma}\check{V}^\mathsf{T}$, in which $\check{U}^\mathsf{T}\check{U} = I_{n_\mathrm{R}}$, $\check{V}^\mathsf{T}\check{V} = I_{n_\mathrm{R}}$, and $\check{\Sigma} = \operatorname{diag}(\sigma_1(V), \dots, \sigma_{n_\mathrm{R}}(V))$. From (5.14) on page 86 we know

$$\begin{aligned}
\|x(i) - Vx_\mathrm{R}(i)\|^2 &= x^\mathsf{T}(i)x(i) - 2x^\mathsf{T}(i)\check{U}\check{\Sigma}\check{V}^\mathsf{T}x_\mathrm{R}(i) + x_\mathrm{R}^\mathsf{T}(i)\check{V}\check{\Sigma}^2\check{V}^\mathsf{T}x_\mathrm{R}(i)\\
&= \left\|\check{U}^\mathsf{T}x(i) - \check{\Sigma}\check{V}^\mathsf{T}x_\mathrm{R}(i)\right\|^2 + \left\|\left(I_n - \check{U}\check{U}^\mathsf{T}\right)x(i)\right\|^2.
\end{aligned}$$

Hence, the constraint (6.10b) can be reformulated.

$$\begin{aligned}
&\alpha\|x(i) - Vx_\mathrm{R}(i)\| \le \bar{\Delta}(i;i)\\
&\iff \alpha\left(\left\|\check{U}^\mathsf{T}x(i) - \check{\Sigma}\check{V}^\mathsf{T}x_\mathrm{R}(i)\right\|^2 + \left\|\left(I_n - \check{U}\check{U}^\mathsf{T}\right)x(i)\right\|^2\right)^{1/2} \le \bar{\Delta}(i;i)\,.
\end{aligned}$$

The second summand $\left\|(I_n - \check{U}\check{U}^\mathsf{T})x(i)\right\| =: p(x)$ is independent of the optimization variables. Thus, it can be computed before the optimization. Altogether, the constraint (6.10b) is equivalent to

$$\iff \alpha\left\|\begin{bmatrix}\check{U}^\mathsf{T}x(i) - \check{\Sigma}\check{V}^\mathsf{T}x_\mathrm{R}(i)\\ p(x)\end{bmatrix}\right\| \le \bar{\Delta}(i;i)\,. \tag{6.12}$$

The dimension of the constraints is reduced by splitting the error into two orthogonal parts. First, the error along the span of the columns of V, which can be influenced by the state of the reduced model. Second, the norm of the orthogonal projection of the state of the preprocessed model onto the kernel of V^T, which is independent of the state of the reduced model.

Altogether, the dimension $n+1$ of the second-order cone constraints (6.10a) and (6.10b) is reduced to the dimension $\tilde{r} + 1 \le n_\mathrm{R} + n_\mathrm{U} + 1$ in (6.11) and the dimension $n_\mathrm{R} + 2$ in (6.12).

We summarize the reformulations by stating the following optimization problem that is equivalent to Problem 6.9.

Problem 6.11 (Convex optimization problem for MPC using the reduced model and error bounding system with low-dimensional constraints).

$$\underset{\bar{u},x_{\mathrm{R}}(i)\in\mathcal{I},\bar{x}_{\mathrm{R}},\bar{\Delta}}{\text{minimize}} \quad J_\Delta\big(x_{\mathrm{R}}(i),\bar{u}(\cdot;i)\big) = \sum_{\tau=i}^{i+T-1} F_{\mathrm{R}}\big(\bar{x}_{\mathrm{R}}(\tau;i),\bar{u}(\tau;i)\big) + E_\Delta\big(\bar{x}_{\mathrm{R}}(i+T;i)\big)$$

subject to

$$\bar{x}_{\mathrm{R}}(\tau+1;i) = A_{\mathrm{R}}\bar{x}_{\mathrm{R}}(\tau;i) + B_{\mathrm{R}}\bar{u}(\tau;i)\,, \tag{6.13a}$$

$$\bar{x}_{\mathrm{R}}(i;i) = x_{\mathrm{R}}(i)\,, \tag{6.13b}$$

$$c_k^{\mathsf{T}}\begin{bmatrix} V\bar{x}_{\mathrm{R}}(\tau;i) \\ \bar{u}(\tau;i)\end{bmatrix} \le d_k - \bar{\Delta}(\tau;i)\left\| c_k^{\mathsf{T}}\begin{bmatrix} I_n \\ 0_{n_{\mathrm{U}}\times n}\end{bmatrix}\right\|, \qquad k=1,\dots,n_{\mathrm{C}}\,, \tag{6.13c}$$

$$\left\| \left(P_\Delta^\Omega\right)^{1/2}\bar{x}_{\mathrm{R}}(i+T;i)\right\| \le \sqrt{\gamma_{\mathrm{x}}}\,, \tag{6.13d}$$

$$0 \le \bar{\Delta}(i+T;i) \le \gamma_\Delta\,, \tag{6.13e}$$

$$\alpha\left\| \tilde{\Sigma}\tilde{V}^{\mathsf{T}}\begin{bmatrix}\bar{x}_{\mathrm{R}}(\tau;i) \\ \bar{u}(\tau;i)\end{bmatrix}\right\| \le \bar{\Delta}(\tau+1;i) - \beta\bar{\Delta}(\tau;i)\,, \tag{6.13f}$$

$$\alpha\left\| \begin{bmatrix}\check{U}^{\mathsf{T}}x(i) - \check{\Sigma}\check{V}^{\mathsf{T}}x_{\mathrm{R}}(i) \\ p(x)\end{bmatrix}\right\| \le \bar{\Delta}(i;i)\,, \tag{6.13g}$$

for all $\tau = i,\dots,i+T-1$.

The terms $\check{U}^{\mathsf{T}}x(i)$ and $p(x) = \left\|\big(I_n - \check{U}\check{U}^{\mathsf{T}}\big)x(i)\right\|$ can be computed in advance. Thus, Problem 6.11 is independent of the dimension of the preprocessed model.

The quadratic objective in Problem 6.11 can be reformulated as second-order cone constraints by introducing the additional optimization variables ζ and $\phi := \begin{bmatrix}\phi_0 & \dots & \phi_{T-1} & \phi_\Omega\end{bmatrix}$.

Problem 6.12 (SOCP for MPC using the reduced model and error bounding system).

$$\underset{\bar{u},x_{\mathrm{R}}(i)\in\mathcal{I},\bar{x}_{\mathrm{R}},\bar{\Delta},\zeta,\phi}{\text{minimize}} \quad \zeta$$

subject to

$$\left\|\begin{bmatrix}\phi_0 & \dots & \phi_{T-1} & \phi_\Omega\end{bmatrix}\right\| \le \zeta\,, \tag{6.14a}$$

$$\left\| \begin{bmatrix} Q_{\mathrm{R}} & S_{\mathrm{R}} \\ S_{\mathrm{R}}^{\mathsf{T}} & R\end{bmatrix}^{1/2}\begin{bmatrix}\bar{x}_{\mathrm{R}}(\tau;i) \\ \bar{u}(\tau;i)\end{bmatrix}\right\| \le \phi_{\tau-i}\,, \tag{6.14b}$$

$$\left\| \left(Q_\Delta^\Omega\right)^{1/2}\bar{x}_{\mathrm{R}}(i+T;i)\right\| \le \phi_\Omega\,, \tag{6.14c}$$

$$\text{(6.13a)} - \text{(6.13g)}\,, \tag{6.14d}$$

for all $\tau = i,\dots,i+T-1$.

Problem 6.12 is an SOCP as defined in [Boyd and Vandenberghe, 2004]. In comparison, the optimization problem for MPC using the preprocessed model is a quadratically constrained quadratic program (QCQP). SOCPs are more general than QCQPs [Lobo et al., 1998]. Nevertheless, SOCPs and QCQPs are closely related [Boyd and Vandenberghe, 2004]. Furthermore, QCQPs are translated into SOCPs in the IBM ILOG CPLEX Optimization Studio [IBM, 2013, Second order cone programming and non PSD]. Thus, if $n \gg n_{\mathrm{R}} + n_{\mathrm{U}}$, we expect that the computational efficiency is increased by using the proposed MPC scheme instead of an MPC scheme relying on the preprocessed model.

In summary, the optimization of the proposed MPC scheme is equivalent to an SOCP that is independent of the dimension of the preprocessed model. The crucial point has been the relaxation of the nonlinear equality constraints for the error bound. Since this relaxation allows the error bound only to increase, we have been able to show the equivalence of the two optimization problems. Afterwards, the dimension of the SOCP constraints has been reduced from $n+1$ to a value smaller than $n_{\mathrm{R}} + n_{\mathrm{U}} + 1$ by using the full-rank factorization.

6.4 Example: Tubular Reactor

To investigate the computational demand of the three MPC schemes in discrete-time the nonisothermal tubular reactor is reused. The optimization problems of the MPC schemes are solved using the IBM ILOG CPLEX Optimization Studio.

The maximal as well as the median computation time at $t = 0\,\mathrm{s}$ and the performance degradation of Δ-MPC for $\Delta T_{\mathrm{in}} = 25\,\mathrm{K}$ for varying order of the reduced model is depicted in Figure 6.1. In the following, we use the reduced model of order 36, since it is a good compromise of performance degradation and computational demand. In continuous-time a reduced model of order 40 is necessary to achieve a similar performance degradation of 1% for $\Delta T_{\mathrm{in}} = 25\,\mathrm{K}$.

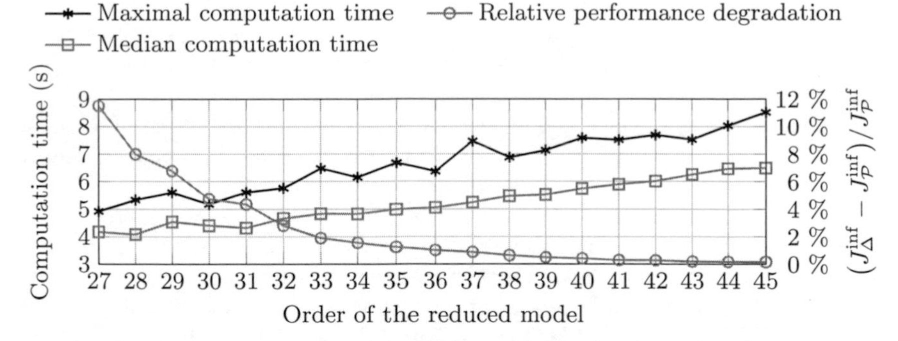

Figure 6.1: Influence of the order of the reduced model on the computation time of Δ-MPC and the relative performance degradation for the plant in closed loop with Δ-MPC compared to $\mathcal{P}$-MPC for $\Delta T_{\mathrm{in}} = 25\,\mathrm{K}$.

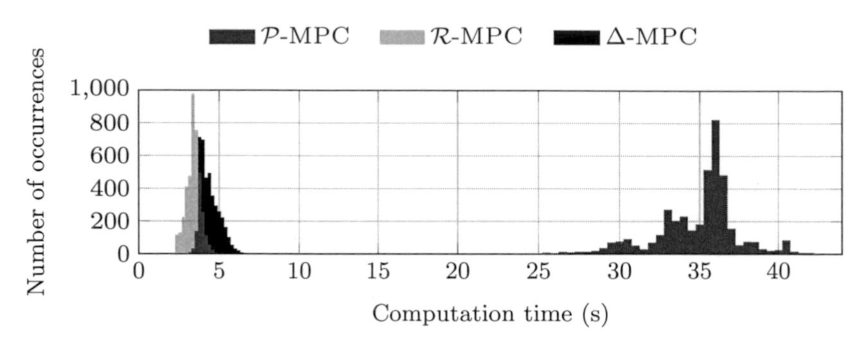

Figure 6.2: Histogram of the computation time of $\mathcal{P}$-MPC, $\mathcal{R}$-MPC, and Δ-MPC for the optimization along the closed-loop trajectories for $t \in [0\,\mathrm{s}, 16\,\mathrm{s}]$.

Table 6.1: Computation time for the optimization along the closed-loop trajectories for $t \in [0\,\mathrm{s}, 16\,\mathrm{s}]$.

	$\mathcal{P}$-MPC	Δ-MPC	$\mathcal{R}$-MPC
Median computation time	35.9 s	4.4 s	3.5 s
Maximal computation time	43.6 s	6.9 s	5.4 s
Median computation time normalized to the value of Δ-MPC	8.23	1	0.80
Maximal computation time normalized to the value of Δ-MPC	6.34	1	0.79

One reason is that in discrete-time the error bound is smaller since it is satisfied only at the sampling instants and not in between as in continuous-time.

A histogram of the computation time of $\mathcal{P}$-MPC, Δ-MPC, and $\mathcal{R}$-MPC for the optimization along the closed-loop trajectories for $t \in [0\,\mathrm{s}, 16\,\mathrm{s}]$ is shown in Figure 6.2.

Furthermore, the maximal and median computation time are stated in Table 6.1. The performance degradation of Δ-MPC compared to $\mathcal{P}$-MPC is almost identical to the continuous-time case plotted in Figure 5.9 on page 107.

The maximal computation time is decreased by Δ-MPC by a factor of 6.34 compared to $\mathcal{P}$-MPC. Furthermore, the $\mathcal{R}$-MPC scheme is only 21% faster than Δ-MPC. Thus, the Δ-MPC scheme shows a sensible compromise between increased computational efficiency and performance degradation. The maximal computation time of Δ-MPC is by a factor of 35 larger than the sampling time of 0.2 s. For the optimization problem of $\mathcal{P}$-MPC without the quadratic terminal constraint a speed up of an order of 100 with no significant decrease in the quality of the model predictive controller is shown in [Wang and Boyd, 2010]. Hence, computing the solution in real time with a specially designed solver is realistic for Δ-MPC.

6.5 Summary

In this chapter, we have shown that for discrete-time systems the optimization of the Δ-MPC scheme is equivalent to an SOCP. Hence, the $\mathcal{P}$-MPC and the Δ-MPC schemes have optimization problems whose classes are closely related. But the dimension of the optimization problem of the Δ-MPC scheme is independent of the dimension of the preprocessed model. Especially, if the dimension of the reduced model plus the dimension of the input is much smaller than the dimension of the plant model, the Δ-MPC scheme is expected to be computationally much more efficient. The computational efficiency of the Δ-MPC scheme in comparison to the $\mathcal{P}$-MPC scheme has been demonstrated by means of the tubular reactor. Furthermore, the Δ-MPC scheme is only slightly less computationally efficient than the $\mathcal{R}$-MPC scheme although the Δ-MPC scheme provides guarantees for the closed-loop behavior.

Chapter 7

Conclusions

To conclude, we summarize and discuss the main results of this thesis followed by an outlook on future research directions.

7.1 Summary

In this thesis, we presented novel methods for model reduction and the utilization of reduced models with the common goal of reduced computational complexity. Since the model reduction error can compromise the application at hand, we provided methods with rigorous guarantees.

In Chapter 3, we introduced an I/O trajectory-based approach for model reduction of nonlinear continuous-time dynamical systems. The approach uses a nonlinear mapping from the state variables of the detailed model to the state variables of the reduced model determined by the observability map. In contrast, many existing model reduction procedures are limited to a linear mapping. The comparison with existing model reduction procedures relying on I/O data in Section 3.4 revealed the novelty of using the observability normal form for model reduction. The formulation as an optimization problem allows for a low complexity functional expression of the reduced model by sparsity enhancing ℓ_1-minimization. Furthermore, the optimization enables to account for additional requirements by constraints on the parameters of the reduced model. We derived necessary and sufficient conditions for local exponential stability of the reduced model and, thereby, extended the approach to preserve the location and local exponential stability of multiple steady states. A computationally tractable model reduction procedure is achieved by utilizing the problem structure to end up with a convex optimization problem. The conditions ensuring the local exponential stability of the reduced model are reformulated with the cone complementarity linearization to deduce an algorithm requiring a sequential convex optimization. The application examples revealed that the trajectory-based model reduction method is able to approximate the relevant I/O behavior with a small number of states demonstrating the potential of the nonlinear mapping between the states of the detailed and the reduced model.

To quantify the uncertainty introduced by the model reduction we generalized an existing a-posteriori error bound in Chapter 4. As a result of adding a decay rate to the norm of the matrix exponential and introducing a preprocessing of the plant model, an asymptotically stable error bounding system can be achieved for every stabilizable system. Moreover, the generalized error bound is considerably tighter, especially for a long simulation time. In contrast to [Löhning et al., 2014], the preprocessing is used instead of an error feedback since it results in a modular approach separating the preprocessing from the error bound and subsequent steps. The generalizations of the error bound can be used in a broad context. One example is dynamic optimization as described in [Hasenauer et al., 2012], which can be used, e.g., for parameter estimation, parameter optimization, and feed-forward control.

We used this error bounding system in Chapter 5 to establish a novel MPC scheme using a reduced model for the prediction to control a possibly unstable high-dimensional linear system. We derived clear design conditions that guarantee satisfaction of hard input and state constraints, recursive feasibility, and asymptotic stability of the equilibrium for the controlled high-dimensional system despite the model reduction error. In contrast, existing results on asymptotic stability of MPC with robustness against the model reduction error either limit the model reduction method to modal truncation, use no error bound, or constant error bounds, which rely on a worst-case analysis. While no error bound prevents the guaranteed satisfaction of hard state constraints, the constant error bounds are used to prove asymptotic stability of a possibly large set around the equilibrium. Furthermore, we showed that the infinite horizon cost functional for the plant is minimized in the proposed MPC scheme under mild assumptions on the terminal set and cost when using the LQR for prestabilization. In this case, the model reduction error is overcome in the cost functional. This leads to a better performance, local optimality around the equilibrium, and a guaranteed upper bound for the cost functional value of the plant. The main novelty of the proposed MPC scheme is the bound of the model reduction error by means of a dynamical system. If the set defining the input and/or state constraints is large in comparison to the initial condition, this error bound is tighter than the constant error bounds relying on a worst-case analysis, which are used in almost all existing results on asymptotic stability of MPC with robustness against the model reduction error. For the tubular reactor, we showed that the error bound of the proposed MPC scheme is less conservative by orders of magnitude than the constant error bound of [Dubljevic et al., 2006]. Although the error bound makes the mentioned guarantees possible, it adds a nonlinear system to the optimization problem. This raises the question of computational complexity. Hence, for discrete-time plant models, we showed that the online optimization problem of the proposed MPC scheme is equivalent to a second-order cone program, which is closely related to the quadratically constrained quadratic programs underlying the MPC scheme using the plant model. Furthermore, we achieved that the dimension of the second-order cone program is independent of the dimension of the plant model. Since the dimension of the reduced model and the input is often much smaller than the order of the plant model, the proposed MPC scheme increases, in general, the computational efficiency. For the tubular reactor, the proposed MPC scheme reduces the maximal computation time by a factor of 6.34 compared to the MPC scheme based on the preprocessed model while achieving the desired region of attraction and resulting in minor differences in the performance. Furthermore, the proposed MPC scheme is only slightly less efficient than the MPC scheme relying only on the reduced model. Thus, the proposed MPC scheme achieves a good trade-off between computational efficiency and conservatism for the tubular reactor while at the same time providing important guarantees for the closed-loop behavior.

7.2 Outlook

The trajectory-based model reduction procedure was used for model reduction of two examples with up to 28 states. Hence, it would be interesting to apply it to more and large-scale detailed models combined with a comparison of other nonlinear model reduction procedures. This allows to further demonstrate the potential of the nonlinear mapping between the states of the detailed and the reduced model. A first step is documented in [Slusarek, 2012], where it is demonstrated that the trajectory-based model reduction is able to approximate a microelectromechanical switch with 240 states with a reduced model with only one state. But for large-scale examples, the order of the reduced model can be limited due to difficulties computing the derivatives of the output of the detailed model. Hence, future research could also incorporate the modulating functions approach, which overcomes the differentiation by using ansatz functions and integration by parts, see [Guo et al., 2016; Rao and Unbehauen, 2006, Section 9.2.1] and the references therein. Furthermore, in order to end up with a convex optimization problem, the objective functional has been changed in Section 3.2.3 such that the short-term prediction error is penalized. Hence, future research should investigate a minimization of the integrated error along the trajectories of the reduced model in order to improve the long-term prediction capabilities of the reduced model. This results in a nonlinear and nonconvex optimization problem. Hence, the initial guess is important in order to find a good solution. One possible initial guess is the solution of the suggested convex approach. A good starting point for this research direction is the optimization of the subspace used for projection by minimizing the integrated error along the trajectories of the reduced model [Bui-Thanh et al., 2007].

For the a-posteriori error bound, several generalizations are possible, which have the potential to increase the tightness. In this thesis, the bound has been generalized to a first-order ODE allowing for only one decay rate. A further generalization could bound the norm of the matrix exponential by the impulse response of a second or higher order ODE. Then, the error bound can be computed by an ODE and the computational demand is only slightly increased. This generalization would result in a flexible bound for the norm of the matrix exponential similar to [Ruiner et al., 2012], but without the high computational demand due to the convolution integral, which has to be reevaluated for all time points. Another starting point to achieve a tighter error bound are the improved error bounds for PDEs presented recently in [Schmidt et al., 2020]. Furthermore, from an application point of view, it can be crucial to have a bound for a linear combination of the states, e.g., a critical constraint in a control problem. One possibility is to add an error bound for a linear functional of the states similar to the approach for second-order systems in [Ruiner et al., 2012]. Another possibility is an improved bound for a linear functional of the states using a dual problem [Haasdonk and Ohlberger, 2011].

The a-posteriori error bound was presented only for LTI systems. Thus, deriving error bounds for nonlinear systems and model reduction methods that are not based on a projection such as the trajectory-based approach are desirable but hard research goals. The a-posteriori error bound can be generalized to parameterized time-varying linear systems with matrices that can be written as sums of scalar continuous functions multiplied with parameter- and time-independent matrices by adapting the bound for the norm of the matrix exponential and the error bounding system similar to [Haasdonk and Ohlberger, 2011]. For nonlinear systems, results on error estimators for kernel systems [Wirtz and Haasdonk, 2012] and model reduction by Galerkin projection combined with the discrete empirical interpolation [Wirtz et al., 2014] exist.

The proposed Δ-MPC approach has one drawback that can compromise the achieved performance. The initial condition of the reduced model is used as decision variable in order to attain recursive feasibility. As a consequence, the trajectory of the reduced model can differ significantly from the trajectory of the plant. Hence, the minimized cost functional does not approximate the cost functional of the plant. A remedy is to choose the LQR as prestabilizing controller. Then, with a proper choice of further design parameters, the Δ-MPC scheme implicitly minimizes the infinite horizon cost functional for the plant as shown in Section 5.5. But limiting the prestabilizing controller to the LQR can noticeably decrease the region of attraction. Hence, further research could improve the performance by minimizing the deviation of the reduced model from the projected initial condition of the plant. Furthermore, the Δ-MPC approach can be extended such that the worst-case cost functional for all possible states of the preprocessed model is minimized similar to [Hasenauer et al., 2012].

The main goal of using a reduced model for MPC is a decreased computational demand. In this thesis, a solver for general second-order cone programs is used in the example in Chapter 6. The optimization problems occurring in MPC possess a structure that can be exploited in order to decrease to computational demand as mentioned, e.g., in [Kögel and Findeisen, 2013; Wang and Boyd, 2010]. Hence, it would be interesting to develop an efficient method to solve the second-order cone program occurring in Δ-MPC. A starting point is the solution method for the quadratically constrained quadratic program occurring in MPC of linear systems [Kögel and Findeisen, 2013].

The main advantage of the Δ-MPC approach is the robustness against the model reduction error. This error is linear in the inputs and states. In contrast, robust MPC often considers an additive bounded disturbance in the plant dynamics, e.g., [Mayne et al., 2005]. Hence, generalization of the Δ-MPC approach to plants with an additive bounded disturbance is a worthwhile research direction. This could be achieved by considering the additive bounded disturbance in the residuum of the a-posteriori error bound. Another uncertainty considered in robust MPC are polytopic uncertain linear time-varying systems, which can be interpreted as a possibly conservative approximation of nonlinear systems, e.g., [Kothare et al., 1996; Kouvaritakis et al., 2000]. Hence, as a first step towards nonlinear systems, the generalization of Δ-MPC to a polytopic uncertain linear time-varying systems is another interesting research direction. Further research opportunities are the combination of the Δ-MPC approach with economic MPC [Angeli et al., 2012; Faulwasser et al., 2018] and output feedback [Dihlmann and Haasdonk, 2016; Kögel and Findeisen, 2015; Köhler et al., 2019; Lorenzetti et al., 2019; Schmidt and Haasdonk, 2016]. For the output feedback one can exploit that not the full state is required for Δ-MPC as stated in Proposition 5.10.

Finally, the results on asymptotic stability and constraint satisfaction for the plant in closed loop with the proposed model predictive controller provide conditions on the reduced model, the model reduction error, and design parameters of the controller. These conditions can be exploited in the future to develop a method for model reduction for constrained linear systems controlled by the developed MPC scheme. Good starting points for this promising research direction are an existing model reduction approach that uses an optimization framework to determine the projection basis [Bui-Thanh et al., 2007] and an approach that considers the accuracy of the reduced model and the extent of the constraint tightening [Choroszucha, 2017].

Appendix A

Linearization and Spatial Discretization of the Tubular Reactor

To apply the a-posteriori error bound and the proposed MPC schemes using a reduced model to the tubular chemical reactor introduced in Section 2.4, an LTI ODE model is required. In Section 2.4.1, we mention that the PDE model of the tubular reactor (2.14)

$$\begin{aligned}\frac{\partial C(t,z)}{\partial t} &= -v\frac{\partial C(t,z)}{\partial z} - k_0\, C(t,z)\exp\left(-\frac{E}{R_{\text{gas}}T(t,z)}\right),\\ \frac{\partial T(t,z)}{\partial t} &= -v\frac{\partial T(t,z)}{\partial z} + G_{\text{r}}\, C(t,z)\exp\left(-\frac{E}{R_{\text{gas}}T(t,z)}\right) + H_{\text{r}}\big(T_{\text{w}}(t,z) - T(t,z)\big)\end{aligned}$$

with the boundary conditions $C(t,0) = C_{\text{in}}$, $T(t,0) = T_{\text{in}}$ and initial conditions $C(0,z) = C_0(z)$, $T(0,z) = T_0(z)$ is linearized and discretized. In this chapter, we present the details, especially the ones required to compute the resulting model (2.15) given in Section 2.4.1. The following procedure and, hence, this chapter are based largely on [Agudelo et al., 2007b].

First, the PDE model is linearized around the nominal inlet fluid temperature, the nominal inlet reactant concentration, the jackets temperatures $T_{\text{J,lin}}$, and the resulting steady state profiles for the reactant concentration $C_{\text{lin}}(\cdot)$ and the fluid temperature $T_{\text{lin}}(\cdot)$. The values of the jacket temperatures $T_{\text{J,lin}} = \begin{bmatrix}374.6\,\text{K} & 310.1\,\text{K} & 325.2\,\text{K}\end{bmatrix}^\top$ are computed in [Agudelo et al., 2007b] by means of a sequential optimization. For details of the optimization procedure we refer to [Agudelo et al., 2007b].

We denote the difference to the steady state profiles with

$$\begin{aligned}C^{\Delta}(t,z) &= C(t,z) - C_{\text{lin}}(z)\,,\\ T^{\Delta}(t,z) &= T(t,z) - T_{\text{lin}}(z)\,,\\ T_{\text{w}}^{\Delta}(t,z) &= \begin{cases} T_{\text{J},1}(t) - T_{\text{J,lin},1} & \text{if } 0 \le z < {}^{1}\!/\!{}_{3}L\,,\\ T_{\text{J},2}(t) - T_{\text{J,lin},2} & \text{if } {}^{1}\!/\!{}_{3}L \le z < {}^{2}\!/\!{}_{3}L\,,\\ T_{\text{J},3}(t) - T_{\text{J,lin},3} & \text{if } {}^{2}\!/\!{}_{3}L \le z \le L\,.\end{cases}\end{aligned}$$

With the abbreviations

$$\begin{aligned}
\alpha_{\mathrm{CC}}(z) &:= k_0 \exp\left(-\frac{E}{R_{\mathrm{gas}} T_{\mathrm{lin}}(z)}\right), \\
\alpha_{\mathrm{CT}}(z) &:= k_0\, C_{\mathrm{lin}}(z) \frac{E}{R_{\mathrm{gas}} T_{\mathrm{lin}}^2(z)} \exp\left(-\frac{E}{R_{\mathrm{gas}} T_{\mathrm{lin}}(z)}\right), \\
\alpha_{\mathrm{TC}}(z) &:= -G_{\mathrm{r}} \exp\left(-\frac{E}{R_{\mathrm{gas}} T_{\mathrm{lin}}(z)}\right), \\
\alpha_{\mathrm{TT}}(z) &:= -G_{\mathrm{r}}\, C_{\mathrm{lin}}(z) \frac{E}{R_{\mathrm{gas}} T_{\mathrm{lin}}^2(z)} \exp\left(-\frac{E}{R_{\mathrm{gas}} T_{\mathrm{lin}}(z)}\right) + H_{\mathrm{r}},
\end{aligned}$$

the linearized PDE model is given by

$$\begin{aligned}
\frac{\partial C^{\Delta}(t,z)}{\partial t} &= -v \frac{\partial C^{\Delta}(t,z)}{\partial z} - \alpha_{\mathrm{CC}}(z) C^{\Delta}(t,z) - \alpha_{\mathrm{CT}}(z) T^{\Delta}(t,z), \\
\frac{\partial T^{\Delta}(t,z)}{\partial t} &= -v \frac{\partial T^{\Delta}(t,z)}{\partial z} - \alpha_{\mathrm{TC}}(z) C^{\Delta}(t,z) - \alpha_{\mathrm{TT}}(z) T^{\Delta}(t,z) + H_{\mathrm{r}} T_{\mathrm{w}}^{\Delta}(t,z)
\end{aligned}$$

with the boundary conditions

$$C^{\Delta}(t,0) = C_{\mathrm{in}} - C_{\mathrm{nom}} \quad \text{and} \quad T^{\Delta}(t,0) = T_{\mathrm{in}} - T_{\mathrm{nom}}$$

and initial conditions

$$C^{\Delta}(0,z) = C_0(z) - C_{\mathrm{nom}} \quad \text{and} \quad T^{\Delta}(0,z) = T_0(z) - T_{\mathrm{nom}}.$$

After the linearization, the reactor is divided into $N = 150$ sections. Then, the partial derivatives with respect to z are approximated by backward differences . This results in the ODE model

$$\begin{aligned}
\dot{C}^{\Delta}(t, i\,\Delta z) =& -\frac{v}{\Delta z}\big(C^{\Delta}(t, i\,\Delta z) - C^{\Delta}(t, (i-1)\Delta z)\big) - \alpha_{\mathrm{CC}}(i\,\Delta z) C^{\Delta}(t, i\,\Delta z) \\
& - \alpha_{\mathrm{CT}}(i\,\Delta z) T^{\Delta}(t, i\,\Delta z), \\
C^{\Delta}(0, i\,\Delta z) =& C_0(i\,\Delta z) - C_{\mathrm{nom}}, \\
\dot{T}^{\Delta}(t, i\,\Delta z) =& -\frac{v}{\Delta z}\big(T^{\Delta}(t, i\,\Delta z) - T^{\Delta}(t, (i-1)\Delta z)\big) - \alpha_{\mathrm{TC}}(i\,\Delta z) C^{\Delta}(t, i\,\Delta z) \\
& - \alpha_{\mathrm{TT}}(i\,\Delta z) T^{\Delta}(t, i\,\Delta z) + H_{\mathrm{r}} T_{\mathrm{w}}^{\Delta}(t, i\,\Delta z), \\
T^{\Delta}(0, i\,\Delta z) =& T_0(i\,\Delta z) - T_{\mathrm{nom}},
\end{aligned}$$

for all $i = 1, \ldots, N$.

To rewrite the ODE model in state space form, we define the vectors for the concentrations and temperatures at the grid points $i\,\Delta z$, $i = 1, \ldots, N$,

$$C_{\mathrm{grid}}(t) := \begin{bmatrix} C(t, \Delta z) \\ \vdots \\ C(t, N\,\Delta z) \end{bmatrix} \in \mathbb{R}^N, \qquad T_{\mathrm{grid}}(t) := \begin{bmatrix} T(t, \Delta z) \\ \vdots \\ T(t, N\,\Delta z) \end{bmatrix} \in \mathbb{R}^N,$$

as well as the state vector and the linearization point

$$x(t) := \begin{bmatrix} \frac{C_{\text{grid}}(t)}{C_{\text{nom}}} \\ \frac{T_{\text{grid}}(t)}{T_{\text{nom}}} \\ \frac{T_{\text{J}}(t)}{T_{\text{nom}}} \end{bmatrix} \in \mathbb{R}^{N+N+3}, \qquad x_{\text{lin}} := \begin{bmatrix} C_{\text{lin}}(\Delta z)/C_{\text{nom}} \\ \vdots \\ C_{\text{lin}}(N\,\Delta z)/C_{\text{nom}} \\ T_{\text{lin}}(\Delta z)/T_{\text{nom}} \\ \vdots \\ T_{\text{lin}}(N\,\Delta z)/T_{\text{nom}} \\ T_{\text{J,lin}}/T_{\text{nom}} \end{bmatrix} \in \mathbb{R}^{N+N+3},$$

and the input vector consisting of the normalized time derivatives of the jacket temperatures

$$u(t) := \dot{T}_{\text{J}}(t)/T_{\text{nom}} \in \mathbb{R}^3 .$$

The time derivatives of the jacket temperatures are considered as inputs instead of the jacket temperatures to allow for a penalization of the time derivatives of the jacket temperatures in the control objective. With the state and input vector, the ODE model can be written as

$$\dot{x}(t) = A\big(x(t) - x_{\text{lin}}\big) + Bu(t) + B_{\text{T}}\big(T_{\text{in}}(t) - T_{\text{nom}}\big)/T_{\text{nom}}\,, \quad x(0) = x_0\,, \tag{A.1}$$

in which

$$A = \begin{bmatrix} A_{\text{CC}} + A_{\text{v}} & A_{\text{CT}} & 0_{N\times 3} \\ A_{\text{TC}} & A_{\text{TT}} + A_{\text{v}} & A_{\text{TJ}} \\ 0_{3\times N} & 0_{3\times N} & 0_{3\times 3} \end{bmatrix} \in \mathbb{R}^{(N+N+3)\times(N+N+3)}, \tag{A.2a}$$

$$A_{\text{v}} = \frac{v}{\Delta z}\left(\begin{bmatrix} 0_{N-1}^{\top} & 0 \\ I_{N-1} & 0_{N-1} \end{bmatrix} - I_N\right) \in \mathbb{R}^{N\times N},$$

$$A_{\text{CC}} = -\operatorname{diag}\big(\alpha_{\text{CC}}(\Delta z), \alpha_{\text{CC}}(2\,\Delta z), \ldots, \alpha_{\text{CC}}(N\,\Delta z)\big) \in \mathbb{R}^{N\times N},$$

$$A_{\text{CT}} = -\frac{T_{\text{nom}}}{C_{\text{nom}}}\operatorname{diag}\big(\alpha_{\text{CT}}(\Delta z), \alpha_{\text{CT}}(2\,\Delta z), \ldots, \alpha_{\text{CT}}(N\,\Delta z)\big) \in \mathbb{R}^{N\times N},$$

$$A_{\text{TC}} = -\frac{C_{\text{nom}}}{T_{\text{nom}}}\operatorname{diag}\big(\alpha_{\text{TC}}(\Delta z), \alpha_{\text{TC}}(2\,\Delta z), \ldots, \alpha_{\text{TC}}(N\,\Delta z)\big) \in \mathbb{R}^{N\times N},$$

$$A_{\text{TT}} = -\operatorname{diag}\big(\alpha_{\text{TT}}(\Delta z), \alpha_{\text{TT}}(2\,\Delta z), \ldots, \alpha_{\text{TT}}(N\,\Delta z)\big) \in \mathbb{R}^{N\times N},$$

$$A_{\text{TJ}} = H_{\text{r}}\operatorname{diag}\big(1_{N/3\times 1}, 1_{N/3\times 1}, 1_{N/3\times 1}\big) \in \mathbb{R}^{N\times 3},$$

$$B = \begin{bmatrix} 0_{2N\times 3} \\ I_3 \end{bmatrix} \in \mathbb{R}^{(2N+3)\times 3}, \tag{A.2b}$$

$$B_{\text{T}} = \frac{v}{\Delta z}\begin{bmatrix} 0_N \\ 1 \\ 0_{N+2} \end{bmatrix} \in \mathbb{R}^{N+1+(N+2)},$$

$$x_0 = \begin{bmatrix} C^{\Delta}(0,\Delta z)/C_{\text{nom}} \\ \vdots \\ C^{\Delta}(0,N\,\Delta z)/C_{\text{nom}} \\ T^{\Delta}(0,\Delta z)/T_{\text{nom}} \\ \vdots \\ T^{\Delta}(0,N\,\Delta z)/T_{\text{nom}} \\ \big(T_{\text{J}}(0) - T_{\text{J,lin}}\big)/T_{\text{nom}} \end{bmatrix} \in \mathbb{R}^{2N+3}.$$

For the computation of A_{CC}, A_{CT}, A_{TC}, and A_{TT} the steady state profiles $C_{\text{lin}}(\cdot)$ and $T_{\text{lin}}(\cdot)$ are required at the grid points $i\,\Delta z$, $i = 1,\ldots,N$. For this purpose, in [Agudelo et al., 2007b], the equations $\partial C(t,z)/\partial t = 0$, $\partial T(t,z)/\partial t = 0$ are inserted into the nonlinear PDE model (2.14). Afterwards the reactor is divided into N sections and the spatial derivatives are approximated by forward differences. This results in the steady state model [Agudelo et al., 2007b, (3)]

$$\begin{aligned} C_{\text{lin}}(0) &= C_{\text{nom}}\,, \\ T_{\text{lin}}(0) &= T_{\text{nom}}\,, \\ C_{\text{lin}}\big((i+1)\Delta z\big) &= C_{\text{lin}}(i\,\Delta z) - \frac{k_0\,\Delta z}{v} C_{\text{lin}}(i\,\Delta z) \exp\left(\frac{-E}{R\,T_{\text{lin}}(i\,\Delta z)}\right), \\ T_{\text{lin}}\big((i+1)\Delta z\big) &= T_{\text{lin}}(i\,\Delta z) + \frac{G_{\text{r}}\Delta z}{v} C_{\text{lin}}(i\,\Delta z) \exp\left(\frac{-E}{R\,T_{\text{lin}}(i\,\Delta z)}\right) \\ &\qquad + \frac{H_{\text{r}}\Delta z}{v}\big(T_{\text{w}}(t,i\,\Delta z) - T_{\text{lin}}(i\,\Delta z)\big)\,, \end{aligned}$$

for all $i = 0,\ldots,N-1$.

Appendix B

Model of the MAPK Cascade

In Section 3.2 the proposed model reduction approach is applied to the model of the mitogen-activated protein kinase (MAPK) cascade. The ODEs of the MAPK cascade model is identical to the model for the case of cooperative inhibition discussed in [Kholodenko, 2000]. The MAPK cascade model is illustrated in Figure B.1. The model has eight state variables, which are the concentration of the three kinases Raf, MEK, and ERK in different phosphorylation states

$$x = \big[[\text{Raf}] \;\; [\text{Raf-P}] \;\; [\text{MEK}] \;\; [\text{MEK-P}] \;\; [\text{MEK-PP}] \;\; [\text{ERK}] \;\; [\text{ERK-P}] \;\; [\text{ERK-PP}]\big]^{\mathsf{T}}.$$

The input of the MAPK cascade is the activity of the enzyme Ras,

$$u_\varphi(t) = c \cdot [\text{Ras}](t)\,,$$

which catalyzes the phosphorylation of Raf. The factor c is a normalization constant ensuring $u_\varphi(t) \in [0,1]$. The output is the concentration of double phosphorylated ERK

$$y_{\mathrm{D}}(t) = [\text{ERK-PP}](t)\,.$$

The activated kinases Raf-P and MEK-PP phosphorylate the kinases MEK and ERK, respectively. We consider the case of cooperative inhibition in which the activated kinase ERK-PP inhibits the phosphorylation of Raf. The reaction rates are modeled by first or second-order Hill kinetics

$$\begin{aligned}
v_1(t) &= \frac{V_1 x_1(t)}{\big(1+(x_8(t)/K_{\mathrm{i}})^2\big)\big(K_{\mathrm{m1}}+x_1(t)\big)} u_\varphi(t)\,, & v_2(t) &= \frac{V_2 x_2(t)}{K_{\mathrm{m2}}+x_2(t)}\,,\\
v_3(t) &= \frac{k_3 x_2(t) x_3(t)}{K_{\mathrm{m3}}+x_3(t)}\,, & v_4(t) &= \frac{k_4 x_2(t) x_4(t)}{K_{\mathrm{m4}}+x_4(t)}\,,\\
v_5(t) &= \frac{V_5 x_5(t)}{K_{\mathrm{m5}}+x_5(t)}\,, & v_6(t) &= \frac{V_6 x_4(t)}{K_{\mathrm{m6}}+x_4(t)}\,,\\
v_7(t) &= \frac{k_7 x_5(t) x_6(t)}{K_{\mathrm{m7}}+x_6(t)}\,, & v_8(t) &= \frac{k_8 x_5(t) x_7(t)}{K_{\mathrm{m8}}+x_7(t)}\,,\\
v_9(t) &= \frac{V_9 x_8(t)}{K_{\mathrm{m9}}+x_8(t)}\,, & v_{10}(t) &= \frac{V_{10} x_7(t)}{K_{\mathrm{m10}}+x_7(t)}\,.
\end{aligned}$$

The parameter values are identical to the ones in [Kholodenko, 2000, Table 2 and caption of Figure 2] and are stated in Table B.1. With the reaction rates the ODEs

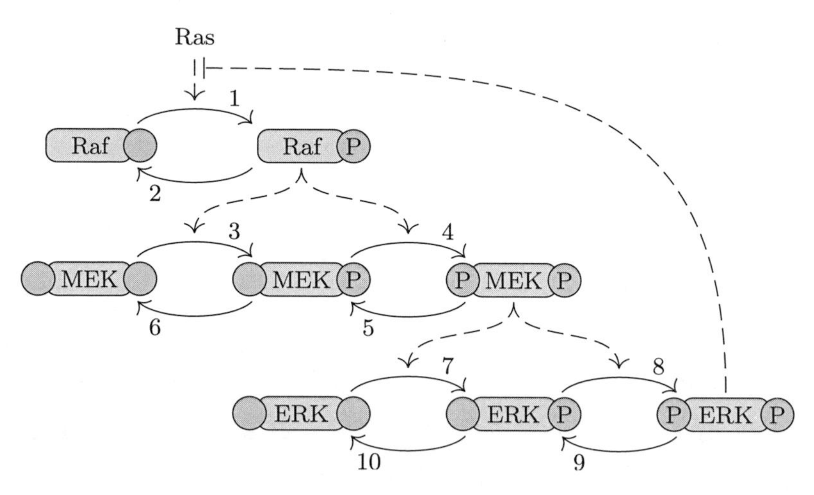

Figure B.1: Illustration of the MAPK cascade model proposed in [Kholodenko, 2000] with conversion reactions (⟶), regulatory interactions (−⤏), and inhibition(−⊣). The schematic is based on [Löhning et al., 2011b].

Table B.1: Parameters of the MAPK cascade [Kholodenko, 2000].

Parameter	Value
V_1	2.5 nmol/s
V_2	0.25 nmol/s
V_5, V_6	0.75 nmol/s
V_9, V_{10}	1.25 nmol/s
k_3, k_4, k_7, k_8	0.025 1/s
K_i	18 nmol
K_m1	50 nmol
K_m2	40 nmol
K_m3, K_m4, K_m5, K_m6, K_m7, K_m8, K_m9, K_m10	100 nmol

of the MAPK cascade are given by

$$
\begin{aligned}
\dot{x}_1(t) &= v_2(t) - v_1(t)\,,\\
\dot{x}_2(t) &= v_1(t) - v_2(t)\,,\\
\dot{x}_3(t) &= v_6(t) - v_3(t)\,,\\
\dot{x}_4(t) &= v_3(t) + v_5(t) - v_4(t) - v_6(t)\,,\\
\dot{x}_5(t) &= v_4(t) - v_5(t)\,,\\
\dot{x}_6(t) &= v_{10}(t) - v_7(t)\,,\\
\dot{x}_7(t) &= v_7(t) + v_9(t) - v_8(t) - v_{10}(t)\,,\\
\dot{x}_8(t) &= v_8(t) - v_9(t)\,,\\
x(0) &= x_0\,,\\
y_{\mathrm{D}}(t) &= x_8(t)\,.
\end{aligned}
$$

The total amount of the three kinases is constant in this model, i.e., $\dot{x}_1(t)+\dot{x}_2(t) = 0$, $\dot{x}_3(t) + \dot{x}_4(t) + \dot{x}_5(t) = 0$, and $\dot{x}_6(t) + \dot{x}_7(t) + \dot{x}_8(t) = 0$. While choosing the initial condition, the total amount of the three kinases

$$
\begin{aligned}
&[\text{Raf}] + [\text{Raf-P}] = 100\,\text{nmol}\,,\\
&[\text{MEK}] + [\text{MEK-P}] + [\text{MEK-PP}] = 300\,\text{nmol}\,,\\
&[\text{ERK}] + [\text{ERK-P}] + [\text{ERK-PP}] = 300\,\text{nmol}\,.
\end{aligned}
$$

has to be considered.

Appendix C

Application of the Model Predictive Control Schemes to a Two-Dimensional Parameterized System

At the beginning of Chapter 5 a two-dimensional example showed that the model reduction error can have a significant influence on the closed-loop behavior of the $\mathcal{R}$-MPC scheme presented in Section 2.3.4. Based on this example, we compare the three MPC schemes $\mathcal{P}$-MPC, $\mathcal{R}$-MPC, and Δ-MPC identified in Table 5.1. To assess the influence of a varying model reduction error, a parameter is added to the two-dimensional system.

This chapter is very similar to [Löhning et al., 2014, Section 7.1]. The main differences originate from the preprocessing and the equivalent cost functional for prestabilization with the LQR.

C.1 Problem Setup

Consider the plant model

$$\dot{x}_{\mathrm{P}}(t) = \begin{bmatrix} 0.1 & 0.75\,\theta \\ 0 & -0.9 \end{bmatrix} x_{\mathrm{P}}(t) + \begin{bmatrix} 0.2 \\ -0.2\,\theta \end{bmatrix} u_{\mathrm{P}}(t)\,, \qquad x_{\mathrm{P}}(0) = x_{\mathrm{P},0}\,,$$

with the parameter $\theta \in [0,1]$. For $\theta = 0.95$, this plant coincides with the plant in Example 5.4. The constraints are $|x_{\mathrm{P},1}(t)| \leq 1$, $|x_{\mathrm{P},2}(t)| \leq 10$, and $|u_{\mathrm{P}}(t)| \leq 0.5$. The stage cost is given by $Q_{\mathrm{P}} = I_2$ and $R_{\mathrm{P}} = 1$. The parameter θ in the state matrix results in a varying influence of the stable mode on the unstable mode. In Section 5.1, we have seen, that $\mathcal{R}$-MPC results in an unstable closed-loop system for $\theta = 0.95$. This shows that the model reduction error is relevant for $\theta = 0.95$ and has to be accounted for in the control design.

C.2 Design of the Model Predictive Controllers

The first step in the design of the Δ-MPC scheme is the preprocessing of the plant. To use the equivalent stage cost, which eliminates the influence of the model reduction error as discussed in Section 5.5, the LQR for the plant denoted with $k^{\mathrm{LQR}}(\theta)$ is used for the prestabilization of the plant. For simplicity, no state transformation is used in the preprocessing. The cost functional for the preprocessed model, i.e., the preprocessed plant model is $J^{\text{inf}}(x_0, u) = \int_0^\infty \|u(t)\|^2 \, dt$.

For the error bounding system an upper bound for the norm of the matrix exponential is required. For simplicity, we set $\alpha = 1$. This results in the decay rate $\beta(\theta) = -0.5\lambda_{\max}\big(A(\theta) + A^{\mathsf{T}}(\theta)\big)$ in which $A(\theta)$ is the state matrix of the preprocessed plant model. For $\theta < 0.992$, the decay rate is positive. One possibility to satisfy Assumption 5.17 for $\theta \in [0.992, 1]$ is to use another prestabilizing feedback.

The reduced model for Δ-MPC is derived from the preprocessed model. Using the matrices $V = W = [1 \quad 0]^{\mathsf{T}}$ as in Example 5.4 for model reduction by projection results in the reduced model

$$\dot{x}_{\mathrm{R}}(t) = \big(0.1 - 0.2k_1^{\mathrm{LQR}}(\theta)\big)x_{\mathrm{R}}(t) + 0.2u(t)\,, \qquad x_{\mathrm{R}}(0) = x_{\mathrm{R},0}\,, \tag{C.1}$$

in which $k_1^{\mathrm{LQR}}(\theta)$ is the first element of the LQR of the plant. The reduced model is asymptotically stable for all $\theta \in [0, 1]$. At $\theta = 0$ the reduced model reflects exactly the mode of the preprocessed model closest to the imaginary axis. With increasing θ, the reduced model reproduces the mode of the preprocessed model closest to the imaginary axis worse. For example, at $\theta = 1$ the reduced model is $\dot{x}_{\mathrm{R}}(t) = -0.75x_{\mathrm{R}}(t) + 0.2u(t)$ and the mode of the preprocessed model closest to the imaginary axis is described by $\dot{x}(t) = -0.1x(t) + 0.05u(t)$. Hence, the model reduction error increases with θ.

The error bounding system is given by

$$\dot{\Delta}(t) = -\beta(\theta)\Delta(t) + 0.2\,\theta\Big|k_1^{\mathrm{LQR}}(\theta)x_{\mathrm{R}}(t) - u(t)\Big|\,, \qquad \Delta(0) = \|x_0 - Vx_{\mathrm{R},0}\|\,.$$

With Proposition 5.7, the constraints for Δ-MPC are $|x_{\mathrm{R}}(t)| \leq 1 - \Delta(t)$, $\Delta(t) \leq 10$, and $\big|u(t) - k_1^{\mathrm{LQR}}(\theta)x_{\mathrm{R}}(t)\big| \leq 0.5 - \big\|k^{\mathrm{LQR}}(\theta)\big\|\Delta(t)$. The constraint $\Delta(t) \leq 10$ can be omitted, since $|x_{\mathrm{R}}(t)| \leq 1 - \Delta(t)$ implies $\Delta(t) \leq 1$.

The $\mathcal{R}$-MPC scheme could also use the reduced model derived from the preprocessed model. But, due to the prestabilizing feedback, also the pure (and dominant) input constraint could not be guaranteed. Hence, as in Example 5.4, the reduced model of the $\mathcal{R}$-MPC scheme is obtained by projection of the plant model with the matrices $V = W = [1 \quad 0]^{\mathsf{T}}$. This results in

$$\dot{x}_{\mathrm{RP}}(t) = 0.1x_{\mathrm{RP}}(t) + 0.2u_{\mathrm{P}}(t)\,, \qquad x_{\mathrm{RP}}(0) = x_{\mathrm{RP},0}\,.$$

Since the reduced model is independent of θ, also the optimization problem of $\mathcal{R}$-MPC is independent of θ. In contrast, the varying influence of the model reduction error is taken into account within Δ-MPC by the error bounding system.

For Δ-MPC, a terminal set of the form (5.21) is used. Since the reduced model in (C.1) is asymptotically stable, the assumptions of Lemma 5.20 are satisfied for $\theta < 0.992$. Since the stage cost does not penalize the state, the unique positive semidefinite solution of the ARE (5.23) is $0_{n_{\mathrm{R}} \times n_{\mathrm{R}}}$. Hence, no terminal controller and no terminal cost is used. Since $n_{\mathrm{R}} = 1$, we do not limit the terminal set by choosing $P^{\Omega}_{\Delta} = 1$. Thus, γ_{x} and γ_{Δ} have to be determined such that (5.25) and (5.26) are satisfied. This results in

$$0.2\,\theta\sqrt{\gamma_{\mathrm{x}}}\left|k_1^{\mathrm{LQR}}(\theta)\right| \leq \beta\gamma_{\Delta}\,, \tag{C.2a}$$

$$\sqrt{\gamma_{\mathrm{x}}} + \gamma_{\Delta} \leq 1\,, \tag{C.2b}$$

$$\sqrt{\gamma_{\mathrm{x}}}\left|k_1^{\mathrm{LQR}}(\theta)\right| + \gamma_{\Delta}\left\|k^{\mathrm{LQR}}(\theta)\right\| \leq 0.5\,. \tag{C.2c}$$

Due to the last two inequalities, there is a compromise between a large terminal set for x_{R} and a large terminal set for Δ. We set γ_{x} to the largest possible value such that there exist a γ_{Δ} satisfying (C.2), as this results in a larger region of attraction for almost all $\theta \in [0,1]$. For $\mathcal{P}$-MPC and $\mathcal{R}$-MPC the terminal set and terminal cost as in Example 5.4 is used.

The three MPC schemes are implemented with a sampling time δ of 0.2, a prediction horizon of $50\,\delta$, and inputs that are piecewise constant within each sampling time.

C.3 Time Response

Trajectories of the closed loop with $\mathcal{P}$-MPC and Δ-MPC are shown in Figure C.1 for $\theta = 0.95$. As initial condition the maximal x_1 for $x_2 = 0$ in the region of attraction of Δ-MPC is taken. The trajectory for the closed loop with $\mathcal{R}$-MPC is depicted in Figure 5.1. Despite the large θ and, hence, large model reduction error, the trajectories of $\mathcal{P}$-MPC and Δ-MPC are almost identical. For $t < 1.6$, the input constraint is active and Δ-MPC uses the input $u_{\Delta}(t)$ to fulfill the constraints. For $t \geq 1.6$, the input $u_{\Delta}(t) \equiv 0$ and, thus, the optimal input $u_{\mathrm{P}}(t) = -k^{\mathrm{LQR}}(\theta)x_{\mathrm{P}}(t)$ is applied as discussed in Section 5.5.

C.4 Asymptotic Stability

The most fundamental distinction between Δ-MPC and $\mathcal{R}$-MPC are the guarantees of Δ-MPC. Figure 5.1 visualizes that $\mathcal{R}$-MPC results in an unstable closed-loop system for $\theta = 0.95$ while Δ-MPC drives the plant to the origin.

C.5 Performance

Figure C.1 suggests that $\mathcal{P}$-MPC and Δ-MPC result in an almost identical performance for this initial condition and $\theta = 0.95$. In the following, we evaluate the performance of the three MPC schemes for varying $x_{\mathrm{P},1}(0)$. For simplicity, we set $x_{\mathrm{P},2}(0) = 0$. This choice does not imply that $x_{\mathrm{P},2}(\cdot) \equiv 0$ as can be seen in Figure C.1.

The performance measure is the infinite horizon cost functional value of the closed loop with the plant model. We denote this closed-loop cost functional value

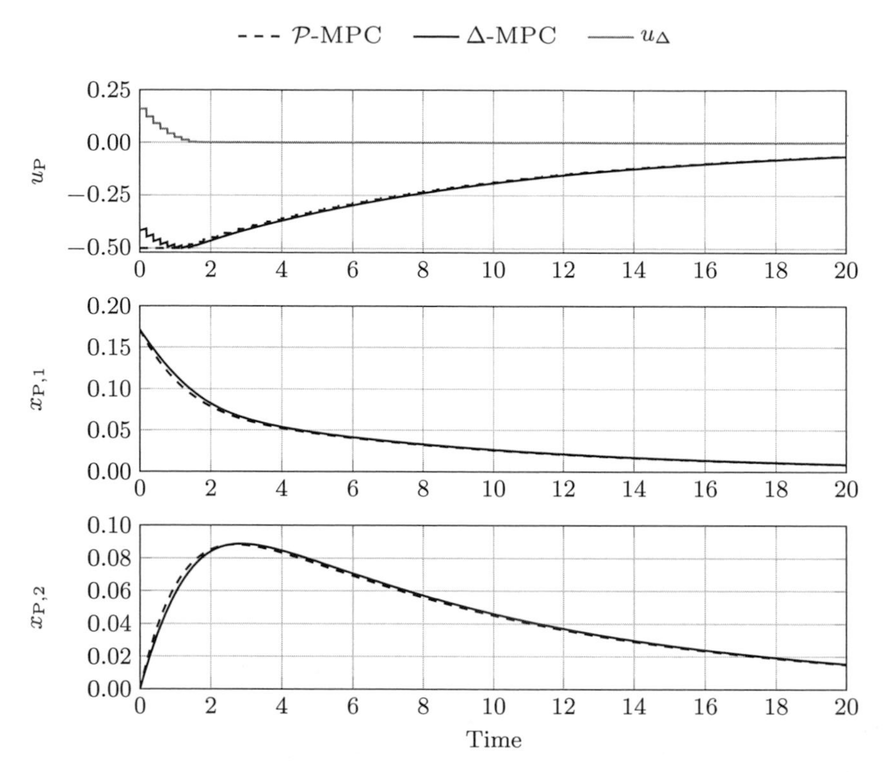

Figure C.1: Trajectories of the plant in closed loop with MPC using the plant model ($\mathcal{P}$-MPC) as well as MPC using the reduced model and error bound (Δ-MPC) for $\theta = 0.95$. To judge the influence of the solution of Δ-MPC to the overall input of the plant, also $u_\Delta(t)$ is shown.

with J_i^{inf} in which $i \in \{\mathcal{P}, \Delta, \mathcal{R}\}$ indicates the utilized MPC scheme. For $\theta \leq 0.5$ and, thus, small influence of the model reduction error, the performance of all three MPC schemes is comparable. However, with increasing θ, the performance of $\mathcal{R}$-MPC gets worse whereas Δ-MPC still achieves a performance comparable to $\mathcal{P}$-MPC, as visualized in Figure C.2a for $\theta = 0.8$.

Figure C.2a also depicts the set of initial conditions with a feasible solutions for $x_{\text{P},2}(0) = 0$. The region of attraction will be considered below. Here, we point out that $\mathcal{R}$-MPC neglects the influence of $x_{\text{P},2}$ and has a feasible solution for all $x_{\text{P},1}(0) \in [-0.75, 0.75]$ independent of θ and $x_{\text{P},2}(0)$. Thus, for $\theta = 0.8$, the set with a feasible solution is even larger than for $\mathcal{P}$-MPC. However, $\mathcal{R}$-MPC is initially feasible and infeasible at a later sampling instant for all $0.52 \leq x_{\text{P},1}(0) \leq 0.75$.

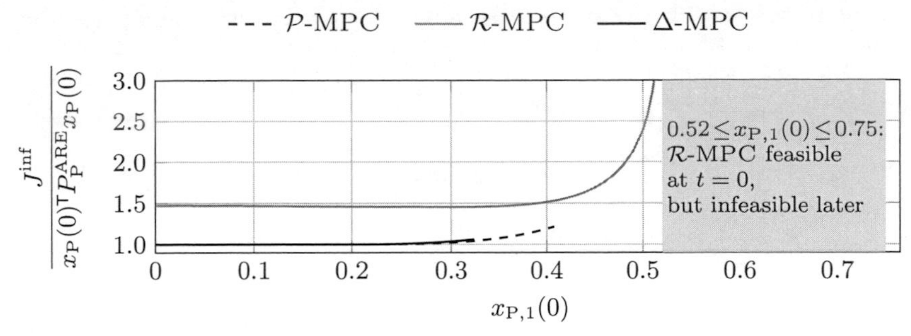

(a) Infinite horizon cost functional values normalized with the optimal cost functional value for the unconstrained plant for $\theta = 0.8$.

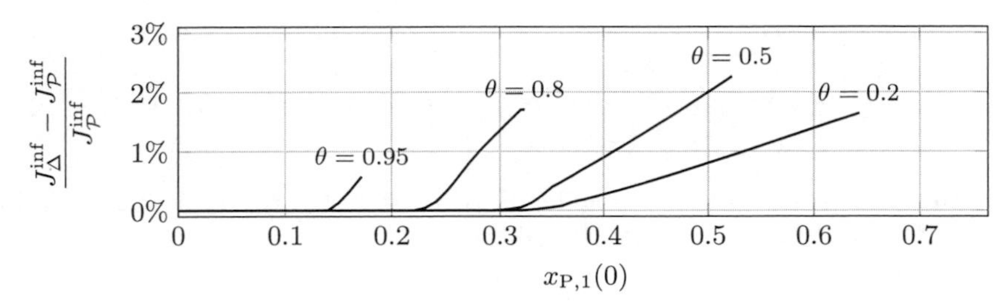

(b) Relative degradation of the infinite horizon cost functional value of Δ-MPC in comparison to $\mathcal{P}$-MPC for $\theta \in \{0.2, 0.5, 0.8, 0.95\}$.

Figure C.2: Infinite horizon cost functional values for the plant in closed loop with $\mathcal{P}$-MPC, $\mathcal{R}$-MPC, and Δ-MPC.

Thus, the larger feasible set of $\mathcal{R}$-MPC is accompanied with the lack of guarantees for the recursive feasibility and asymptotic stability.

In Section 5.5, we have seen that the Δ-MPC scheme is locally optimal when using the equivalent stage cost. To illustrate, how large the region with optimal performance is, the relative performance degradation of Δ-MPC in comparison to $\mathcal{P}$-MPC is depicted in Figure C.2b. The region where Δ-MPC is optimal is reasonably large and the performance degradation is lower than 3%.

C.6 Region of Attraction

The size of the set of initial conditions with guaranteed asymptotic stability is another possibility to assess an MPC scheme. For $\mathcal{P}$-MPC and Δ-MPC this set is given by the set of initial conditions with a feasible solution denoted by $\mathcal{F}_{\mathcal{P}}$ and $\mathcal{F}_{\Delta}$, respectively. The $\mathcal{R}$-MPC scheme gives no guarantee for asymptotic stability and looking at the feasible set can be misleading as indicated in Figure C.2a. Hence, for $\mathcal{R}$-MPC the set of initial conditions with a feasible solution is not evaluated.

To simplify visualization, we restrict $x_{\mathrm{P},2}(0)$ to 0. We denote the length of the interval for $x_{\mathrm{P},1}(0)$ with a feasible solution by

$$l_i := \operatorname*{argmax}_{x_1 \in \mathbb{R}} \left(\begin{bmatrix} x_1 \\ 0 \end{bmatrix} \in \mathcal{F}_i \right) - \operatorname*{argmin}_{x_1 \in \mathbb{R}} \left(\begin{bmatrix} x_1 \\ 0 \end{bmatrix} \in \mathcal{F}_i \right), \qquad i \in \{\mathcal{P}, \Delta\}.$$

The ratio $l_\Delta / l_\mathcal{P}$ is depicted in Figure C.3.

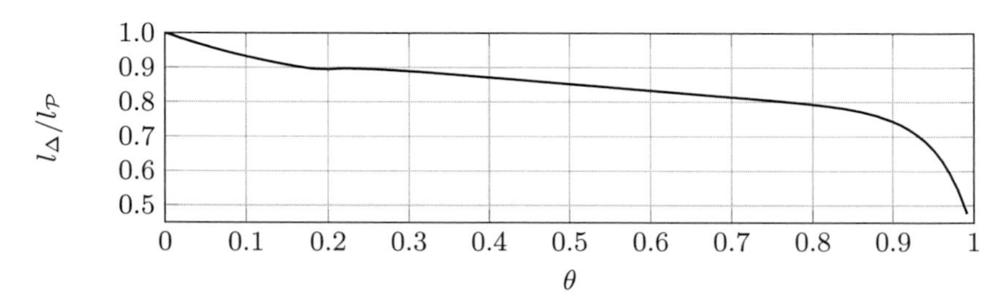

Figure C.3: Ratio of the interval for $x_{\mathrm{P},1}(0)$ within the region of attraction for Δ-MPC and $\mathcal{P}$-MPC. The initial condition of $x_{\mathrm{P},2}$ is set to 0.

For small θ, the residual and thus the error bound is small. Consequently, $l_\Delta \approx l_\mathcal{P}$. With increasing θ the plant is harder to control and both, l_Δ and $l_\mathcal{P}$, decrease. Notably, for $\theta \leq 0.76$ the feasible set for $x_{\mathrm{P},1}(0)$ of Δ-MPC contains 80 % of the feasible set of $\mathcal{P}$-MPC. Even for $\theta = 0.9$, where $\mathcal{R}$-MPC shows a poor performance, the ratio $l_\Delta / l_\mathcal{P}$ is 74 %. For higher values of θ the value of l_Δ decreases substantially since β decreases leading to a slow decay of the error bound. For $\theta \geq 0.992$, Assumption 5.17 is not fulfilled for $\alpha = 1$ and prestabilization with the LQR. For prestabilizing controllers that differ from the LQR, Assumption 5.17 can be satisfied for all $\theta \in [0, 1]$. Furthermore, the feasible set increases noticeably for $\theta \leq 0.3$ and $\theta \geq 0.9$ when the prestabilizing feedback is chosen accordingly.

Summarizing, we have studied a parameter dependent two-dimensional example, in which the parameter changes the influence of the model reduction error. We have compared asymptotic stability, performance, and region of attraction for the three MPC schemes in dependence of this parameter. Even for a significant model reduction error, the Δ-MPC scheme shows a good performance while guaranteeing asymptotic stability.

Bibliography

Advanpix LLC. Multiprecision Computing Toolbox for MATLAB, version 4.3.2.12144. http://www.advanpix.com/, 2017.

D. W. Agar, M. Bertau, M. Busch, M. Casapu, P. Claus, K. H. Delgado, D. Demtröder, O. Deutschmann, R. Dittmeyer, C. Dreiser, D. Eckes, B. Etzold, G. Fieg, H. Freund, J.-D. Grunwaldt, R. Güttel, E. von Harbou, T. Herrmann, K.-O. Hinrichsen, J. Khinast, E. Klemm, N. Kockmann, U. Krewer, M. Nilles, H. Marschall, M. Muhler, S. Palkovits, M. Paul, K. Pflug, S. Radl, J. Sauer, S. Schunk, A. Seidel-Morgenstern, M. Schlüter, M. Schubert, K. Sundmacher, T. Turek, I. Vittorias, H. Vogel, O. Wachsen, H.-W. Zanthoff, and D. Ziegenbalg. Roadmap chemical reaction engineering: an initiative of the ProcessNet subject division chemical reaction engineering. Technical report, DECHEMA e.V., Frankfurt, Germany, 2017.

O. M. Agudelo. *The Application of Proper Orthogonal Decomposition to the Control of Tubular Reactors.* PhD thesis, KU Leuven, Leuven, Belgium, 2009.

O. M. Agudelo, J. J. Espinosa, and B. De Moor. POD-based predictive controller with temperature constraints for a tubular reactor. In *Proceedings of the 46th IEEE Conference on Decision and Control*, pages 3537–3542, New Orleans, LA, USA, 2007a.

O. M. Agudelo, J. J. Espinosa, and B. De Moor. Control of a tubular chemical reactor by means of POD and predictive control techniques. In *Proceedings of the European Control Conference*, pages 1046–1053, Kos, Greece, 2007b.

D. Angeli. A Lyapunov approach to incremental stability properties. *IEEE Transactions on Automatic Control*, 47(3):410–421, 2002.

D. Angeli, R. Amrit, and J. B. Rawlings. On average performance and stability of economic model predictive control. *IEEE Transactions on Automatic Control*, 57(7):1615–1626, 2012.

A. C. Antoulas. An overview of approximation methods for large-scale dynamical systems. *Annual Reviews in Control*, 29:181–190, 2005a.

A. C. Antoulas. *Approximation of Large-Scale Dynamical Systems.* Advances in Design and Control. Society for Industrial and Applied Mathematics, Philadelphia, PA, USA, 2005b.

A. C. Antoulas, P. Benner, and L. Feng. Model reduction by iterative error system approximation. *Mathematical and Computer Modelling of Dynamical Systems*, pages 1–16, 2018.

A. Astolfi. Model reduction by moment matching for linear and nonlinear systems. *IEEE Transactions on Automatic Control*, 55(10):2321–2336, 2010.

L. S. Balasubramhanya and F. J. Doyle III. Nonlinear model-based control of a batch reactive distillation column. *Journal of Process Control*, 10:209–218, 2000.

M. Barrault, Y. Maday, N. C. Nguyen, and A. T. Patera. An empirical interpolation method: application to efficient reduced-basis discretization of partial differential equations. *Comptes Rendus Mathematique*, 339(9):667–672, 2004.

T. Bäthge, S. Lucia, and R. Findeisen. Exploiting models of different granularity in robust predictive control. In *Proceedings of the 55th IEEE Conference on Decision and Control*, pages 2763–2768, Las Vegas, NV, USA, 2016.

U. Baur, P. Benner, and L. Feng. Model order reduction for linear and nonlinear systems: a system-theoretic perspective. *Archives of Computational Methods in Engineering*, 21(4):331–358, 2014.

A. Bemporad and M. Morari. Robust model predictive control: A survey. In A. Garulli, A. Tesi, and A. Vicino, editors, *Robustness in Identification and Control*, Lecture Notes in Control and Information Sciences 245, pages 207–226. Springer-Verlag London, 1999.

P. Benner, S. Gugercin, and K. Willcox. A survey of projection-based model reduction methods for parametric dynamical systems. *SIAM Review*, 57(4): 483–531, 2015.

P. Benner, A. Cohen, M. Ohlberger, and K. Willcox, editors. *Model Reduction and Approximation: Theory and Algorithms*. Computational Science & Engineering. Society for Industrial and Applied Mathematics, Philadelphia, PA, USA, 2017.

D. S. Bernstein. *Matrix Mathematics: Theory, Facts, and Formulas*. Princeton University Press, Princeton, NJ, USA, second edition, 2009.

B. Besselink. *Model Reduction for Nonlinear Control Systems: With Stability Preservation and Error Bounds*. PhD thesis, Eindhoven University of Technology, Eindhoven, Netherlands, 2012.

B. Besselink, U. Tabak, A. Lutowska, N. van de Wouw, H. Nijmeijer, D. J. Rixen, M. E. Hochstenbach, and W. H. A. Schilders. A comparison of model reduction techniques from structural dynamics, numerical mathematics and systems and control. *Journal of Sound and Vibration*, 332(19):4403–4422, 2013.

F. Blanchini. Set invariance in control. *Automatica*, 35(11):1747–1767, 1999.

B. N. Bond and L. Daniel. Stable reduced models for nonlinear descriptor systems through piecewise-linear approximation and projection. *IEEE Transactions on Computer-Aided Design of Integrated Circuits and Systems*, 28(10):1467–1480, 2009.

B. N. Bond, Z. Mahmood, Y. Li, R. Sredojević, A. Megretski, V. Stojanović, Y. Avniel, and L. Daniel. Compact modeling of nonlinear analog circuits using system identification via semidefinite programming and incremental stability certification. *IEEE Transactions on Computer-Aided Design of Integrated Circuits and Systems*, 29(8):1149–1162, 2010.

S. Boyd and L. Vandenberghe. *Convex Optimization.* Cambridge University Press, Cambridge, UK, 2004.

S. Boyd, L. El Ghaoui, E. Feron, and V. Balakrishnan. *Linear Matrix Inequalities in System and Control Theory.* SIAM Studies in Applied Mathematics. Society for Industrial and Applied Mathematics, Philadelphia, PA, USA, 1994.

T. Bui-Thanh, K. Willcox, O. Ghattas, and B. van Bloemen Waanders. Goal-oriented, model-constrained optimization for reduction of large-scale systems. *Journal of Computational Physics*, 224:880–896, 2007.

R. Cagienard, P. Grieder, E. C. Kerrigan, and M. Morari. Move blocking strategies in receding horizon control. *Journal of Process Control*, 17(6):563–570, 2007.

E. F. Camacho and C. Bordons. *Model Predictive Control.* Advanced Textbooks in Control and Signal Processing. Springer-Verlag London, second edition, 2007.

E. J. Candès, J. Romberg, and T. Tao. Robust uncertainty principles: exact signal reconstruction from highly incomplete frequency information. *IEEE Transactions on Information Theory*, 52(2):489–509, 2006.

S. Chaturantabut. *Nonlinear Model Reduction via Discrete Empirical Interpolation.* PhD thesis, Rice University, Houston, TX, USA, 2011.

S. Chaturantabut and D. C. Sorensen. Nonlinear model reduction via discrete empirical interpolation. *SIAM Journal on Scientific Computing*, 32(5):2737–2764, 2010.

C. C. Chen and L. Shaw. On receding horizon feedback control. *Automatica*, 18 (3):349–352, 1982.

H. Chen. *Stability and Robustness Considerations in Nonlinear Model Predictive Control.* PhD thesis, University of Stuttgart, Stuttgart, Germany, 1997.

H. Chen and F. Allgöwer. A quasi-infinite horizon nonlinear model predictive control scheme with guaranteed stability. *Automatica*, 34(10):1205–1217, 1998.

L. Chisci, A. Lombardi, and E. Mosca. Dual-receding horizon control of constrained discrete time systems. *European Journal of Control*, 2(4):278–285, 1996.

L. Chisci, J. A. Rossiter, and G. Zappa. Systems with persistent disturbances: predictive control with restricted constraints. *Automatica*, 37(7):1019–1028, 2001.

D. Chmielewski and V. Manousiouthakis. On constrained infinite-time linear quadratic optimal control. In *Proceedings of the 35th IEEE Conference on Decision and Control*, pages 1319–1324, Kobe, Japan, 1996.

R. B. Choroszucha. *Control and Estimation Oriented Model Order Reduction for Linear and Nonlinear Systems.* PhD thesis, University of Michigan, Ann Arbor, MI, USA, 2017.

G. De Nicolao, L. Magni, and R. Scattolini. Stabilizing receding-horizon control of nonlinear time-varying systems. *IEEE Transactions on Automatic Control*, 43 (7):1030–1036, 1998.

S. L. de Oliveira Kothare and M. Morari. Contractive model predictive control for constrained nonlinear systems. *IEEE Transactions on Automatic Control*, 45(6): 1053–1071, 2000.

U. Desai and D. Pal. A transformation approach to stochastic model reduction. *IEEE Transactions on Automatic Control*, 29(12):1097–1100, 1984.

C. A. Desoer and M. Vidyasagar. *Feedback Systems: Input-Output Properties.* Classics in Applied Mathematics. Society for Industrial and Applied Mathematics, Philadelphia, PA, USA, 2009.

M. Dihlmann and B. Haasdonk. A reduced basis Kalman filter for parametrized partial differential equations. *ESAIM: Control, Optimisation and Calculus of Variations*, 22(3):625–669, 2016.

D. L. Donoho. Compressed sensing. *IEEE Transactions on Information Theory*, 52:1289–1306, 2006.

M. Drohmann, B. Haasdonk, S. Kaulmann, and M. Ohlberger. A software framework for reduced basis methods using Dune-RB and RBmatlab. In A. Dedner, B. Flemisch, and R. Klöfkorn, editors, *Advances in DUNE: Proceedings of the DUNE User Meeting, Held in October 6th–8th 2010 in Stuttgart, Germany*, pages 77–88. Springer-Verlag Berlin Heidelberg, 2012a.

M. Drohmann, B. Haasdonk, and M. Ohlberger. Reduced basis approximation for nonlinear parametrized evolution equations based on empirical operator interpolation. *SIAM Journal on Scientific Computing*, 34(2):A937–A969, 2012b.

S. Dubljevic, N. H. El-Farra, P. Mhaskar, and P. D. Christofides. Predictive control of parabolic PDEs with state and control constraints. *International Journal of Robust and Nonlinear Control*, 16:749–772, 2006.

J. Dunn. On $\mathcal{L}_2$ sufficient conditions and the gradient projection method for optimal control problems. *SIAM Journal on Control and Optimization*, 34(4): 1270–1290, 1996.

L. El Ghaoui, F. Oustry, and M. A. Rami. A cone complementarity linearization algorithm for static output-feedback and related problems. *IEEE Transactions on Automatic Control*, 42(8):1171–1176, 1997.

D. F. Enns. Model reduction with balanced realizations: An error bound and a frequency weighted generalization. In *Proceedings of the 23rd IEEE Conference on Decision and Control*, pages 127–132, Las Vegas, NV, USA, 1984.

T. Faulwasser, L. Grüne, and M. A. Müller. Economic nonlinear model predictive control. *Foundations and Trends® in Systems and Control*, 5(1):1–98, 2018.

P. Feldmann and R. W. Freund. Efficient linear circuit analysis by Padé approximation via the Lanczos process. *IEEE Transactions on Computer-Aided Design of Integrated Circuits and Systems*, 14(5):639–649, 1995.

L. Feng, A. C. Antoulas, and P. Benner. Some a posteriori error bounds for reduced-order modelling of (non-) parametrized linear systems. *Mathematical Modelling and Numerical Analysis*, 51(6):2127–2158, 2017.

E. Fermi, J. Pasta, and S. Ulam. Studies of nonlinear problems. Technical report, Los Alamos report LA-1940, 1955.

R. Findeisen, L. Imsland, F. Allgöwer, and B. A. Foss. State and output feedback nonlinear model predictive control: An overview. *European Journal of Control*, 9:179–195, 2003.

M. Fjeld and B. Ursin. Approximate lumped models of a tubular chemical reactor, and their use in feedback and feedforward control. In *Proceedings of the 2nd IFAC Symposium on Multivariable Technical Control Systems*, pages 1–18, Düsseldorf, Germany, 1971.

H. S. Fogler. *Elements of Chemical Reaction Engineering*. Prentice Hall International Series in the Physical and Chemical Engineering Sciences. Prentice Hall, Upper Saddle River, NJ, USA, fourth edition, 2009.

F. A. C. C. Fontes. A general framework to design stabilizing nonlinear model predictive controllers. *Systems & Control Letters*, 42(2):127–143, 2001.

R. W. Freund. Model reduction methods based on Krylov subspaces. *Acta Numerica*, 12:267–319, 2003.

J. B. Froisy. Model predictive control – building a bridge between theory and practice. *Computers and Chemical Engineering*, 30:1426–1435, 2006.

K. Fujimoto and J. M. A. Scherpen. Balanced realization and model order reduction for nonlinear systems based on singular value analysis. *SIAM Journal on Control and Optimization*, 48(7):4591–4623, 2010.

K. Glover. All optimal Hankel-norm approximations of linear multivariable systems and their $\mathcal{L}_\infty$-error bounds. *International Journal of Control*, 39(6):1115–1193, 1984.

K. Graichen and B. Käpernick. A real-time gradient method for nonlinear model predictive control. In T. Zheng, editor, *Frontiers of Model Predictive Control*. InTech, 2012.

G. Grimm, M. J. Messina, S. E. Tuna, and A. R. Teel. Model predictive control: for want of a local control Lyapunov function, all is not lost. *IEEE Transactions on Automatic Control*, 50(5):546–558, 2005.

E. J. Grimme. *Krylov Projection Methods for Model Reduction*. PhD thesis, University of Illinois at Urbana–Champaign, Urbana–Champaign, IL, USA, 1997.

L. Grüne. Analysis and design of unconstrained nonlinear MPC schemes for finite and infinite dimensional systems. *SIAM Journal on Control and Optimization*, 48(2):1206–1228, 2009.

L. Grüne and J. Pannek. *Nonlinear Model Predictive Control: Theory and Algorithms*. Communications and Control Engineering. Springer International Publishing, second edition, 2017.

L. Grüne and A. Rantzer. On the infinite horizon performance of receding horizon controllers. *IEEE Transactions on Automatic Control*, 53(9):2100–2111, 2008.

D. Grunert, J. Fehr, and B. Haasdonk. Well-scaled, a-posteriori error estimation for model order reduction of large second-order mechanical systems. *ZAMM-Journal of Applied Mathematics and Mechanics / Zeitschrift für Angewandte Mathematik und Mechanik*, 100(8), 2020.

C. Gu. QLMOR: A new projection-based approach for nonlinear model order reduction. In *Proceedings of the 2009 International Conference on Computer-Aided Design*, pages 389–396, San Jose, CA, USA, 2009.

C. Gu. *Model Order Reduction of Nonlinear Dynamical Systems.* PhD thesis, University of California, Berkeley, Berkeley, CA, USA, 2011.

S. Gugercin and A. C. Antoulas. A survey of model reduction by balanced truncation and some new results. *International Journal of Control*, 77(8):748–766, 2004.

Y. Guo, L. Z. Guo, S. A. Billings, and H.-L. Wei. Identification of continuous-time models for nonlinear dynamic systems from discrete data. *International Journal of Systems Science*, 47(12):3044–3054, 2016.

B. Haasdonk and M. Ohlberger. Reduced basis method for finite volume approximations of parametrized linear evolution equations. *ESAIM: Mathematical Modelling and Numerical Analysis*, 42(02):277–302, 2008.

B. Haasdonk and M. Ohlberger. Efficient reduced models and a posteriori error estimation for parametrized dynamical systems by offline/online decomposition. *Mathematical and Computer Modelling of Dynamical Systems*, 17:145–161, 2011.

B. Haasdonk, M. Ohlberger, and G. Rozza. A reduced basis method for evolution schemes with parameter-dependent explicit operators. *ETNA, Electronic Transactions on Numerical Analysis*, 32:145–161, 2008.

J. Hasenauer, M. Löhning, M. Khammash, and F. Allgöwer. Dynamical optimization using reduced order models: A method to guarantee performance. *Journal of Process Control*, 22:1490–1501, 2012.

M. Heinkenschloss, T. Reis, and A. C. Antoulas. Balanced truncation model reduction for systems with inhomogeneous initial conditions. *Automatica*, 47(3): 559–564, 2011.

A. C. Hindmarsh, P. N. Brown, K. E. Grant, S. L. Lee, R. Serban, D. E. Shumaker, and C. S. Woodward. SUNDIALS: Suite of nonlinear and differential/algebraic equation solvers. *ACM Transactions on Mathematical Software*, 31(3):363–396, 2005.

D. Hinrichsen and A. J. Pritchard. *Mathematical Systems Theory I: Modelling, State Space Analysis, Stability and Robustness.* Texts in Applied Mathematics. Springer-Verlag Berlin Heidelberg, 2005.

D. Hinrichsen, E. Plischke, and F. Wirth. State feedback stabilization with guaranteed transient bounds. In *Proceedings of the 15th International Symposium on Mathematical Theory of Networks and Systems*, pages 1–10, South Bend, IN, USA, 2002.

P. Holmes, J. L. Lumley, and G. Berkooz. *Turbulence, Coherent Structures, Dynamical Systems and Symmetry.* Cambridge Monographs on Mechanics. Cambridge University Press, Cambridge, UK, 1996.

S. Hovland and J. T. Gravdahl. Complexity reduction in explicit MPC through model reduction. In *Proceedings of the 17th IFAC World Congress*, pages 7711–7716, Seoul, Korea, 2008.

S. Hovland, J. T. Gravdahl, and K. Willcox. Explicit model predictive control for large-scale systems via model reduction. *Journal of Guidance, Control, and Dynamics*, 31(4):918–926, 2008a.

S. Hovland, C. Løvaas, J. T. Gravdahl, and G. C. Goodwin. Stability of model predictive control based on reduced-order models. In *Proceedings of the 47th IEEE Conference on Decision and Control*, pages 4067–4072, Cancun, Mexico, 2008b.

L. Huisman and S. Weiland. Identification and model predictive control of an industrial glass feeder. In *Proceedings of the 13th IFAC Symposium on System Identification*, pages 1685–1689, Rotterdam, The Netherlands, 2003.

IBM ILOG CPLEX Optimization Studio V12.5.1, User's Manual for CPLEX. IBM, 2013.

T. C. Ionescu and A. Astolfi. Nonlinear moment matching-based model order reduction. *IEEE Transactions on Automatic Control*, 61(10):2837–2847, 2016.

A. Jadbabaie and J. Hauser. On the stability of receding horizon control with a general terminal cost. *IEEE Transactions on Automatic Control*, 50(5):674–678, 2005.

F. Jarmolowitz, D. Abel, T. Wada, and N. Peters. Control of a homogeneous stirred reactor: Trajectory piecewise-linear model for NMPC. In *Proceedings of the European Control Conference*, pages 2301–2306, Budapest, Hungary, 2009.

S. Joe Qin and T. A. Badgwell. A survey of industrial model predictive control technology. *Control Engineering Practice*, 11(7):733–764, 2003.

T. A. Johansen. Reduced explicit constrained linear quadratic regulators. *IEEE Transactions on Automatic Control*, 48(5):823–829, 2003.

Y. Kawano and J. M. A. Scherpen. Empirical differential balancing for nonlinear systems. In *Proceedings of the 20th IFAC World Congress*, pages 6326–6331, Toulouse, France, 2017.

S. S. Keerthi and E. G. Gilbert. Optimal infinite-horizon feedback laws for a general class of constrained discrete-time systems: Stability and moving-horizon approximations. *Journal of Optimization Theory and Applications*, 57(2):265–293, 1988.

C. Kenney and G. Hewer. Necessary and sufficient conditions for balancing unstable systems. *IEEE Transactions on Automatic Control*, 32(2):157–160, 1987.

H. K. Khalil. *Nonlinear Systems.* Prentice Hall, Upper Saddle River, NJ, USA, second edition, 1996.

B. N. Kholodenko. Negative feedback and ultrasensitivity can bring about oscillations in the mitogen-activated protein kinase cascades. *European Journal of Biochemistry*, 267(6):1583–1588, 2000.

M. Kögel and R. Findeisen. On efficient predictive control of linear systems subject to quadratic constraints using condensed, structure-exploiting interior point methods. In *Proceedings of the European Control Conference*, pages 27–34, Zürich, Switzerland, 2013.

M. Kögel and R. Findeisen. Robust output feedback model predictive control using reduced order models. In *Proceedings of the 9th IFAC Symposium on Advanced Control of Chemical Processes*, pages 1008–1014, Whistler, British Columbia, Canada, 2015.

J. Köhler, F. Allgöwer, and M. A. Müller. A simple framework for nonlinear robust output-feedback MPC. In *Proceedings of the 18th European Control Conference*, pages 793–798, 2019.

M. V. Kothare, V. Balakrishnan, and M. Morari. Robust constrained model predictive control using linear matrix inequalities. *Automatica*, 32(10):1361–1379, 1996.

B. Kouvaritakis and M. Cannon. *Model Predictive Control: Classical, Robust and Stochastic.* Advanced Textbooks in Control and Signal Processing. Springer International Publishing, 2016.

B. Kouvaritakis, J. A. Rossiter, and J. Schuurmans. Efficient robust predictive control. *IEEE Transactions on Automatic Control*, 45(8):1545–1549, 2000.

B. Kouvaritakis, M. Cannon, and J. A. Rossiter. Who needs QP for linear MPC anyway? *Automatica*, 38(5):879–884, 2002.

K. Kunisch and S. Volkwein. Proper orthogonal decomposition for optimality systems. *Mathematical Modelling and Numerical Analysis*, 42:1–23, 2008.

H. Kwakernaak and R. Sivan. *Linear Optimal Control Systems.* John Wiley & Sons, Inc., New York, NY, USA, 1972.

S. Lall, J. E. Marsden, and S. Glavaški. A subspace approach to balanced truncation for model reduction of nonlinear control systems. *International Journal of Robust and Nonlinear Control*, 12(6):519–535, 2002.

L. Ljung. *System Identification: Theory for the User.* Prentice Hall Information and System Sciences Series. Prentice Hall, Upper Saddle River, NJ, USA, second edition, 1999.

M. S. Lobo, L. Vandenberghe, S. Boyd, and H. Lebret. Applications of second-order cone programming. *Linear Algebra and its Applications*, 284(1):193–228, 1998.

J. Löfberg. Dualize it: software for automatic primal and dual conversions of conic programs. *Optimization Methods and Software*, 24:313–325, 2009.

F. Logist, I. Y. Smets, and J. F. Van Impe. Optimal control of dispersive tubular chemical reactors: Part I. In *Proceedings of the 16th IFAC World Congress*, Prague, Czech Republic, 2005.

B. Lohmann. Ordnungsreduktion und Dominanzanalyse nichtlinearer Systeme. *at - Automatisierungstechnik*, 42:466–474, 1994.

M. Löhning, J. Hasenauer, and F. Allgöwer. Trajectory-based model reduction of nonlinear biochemical networks employing the observability normal form. In *Proceedings of the 18th IFAC World Congress*, pages 10442–10447, Milano, Italy, 2011a.

M. Löhning, J. Hasenauer, and F. Allgöwer. Steady state stability preserving nonlinear model reduction using sequential convex optimization. In *Proceedings of the 50th IEEE Conference on Decision and Control and European Control Conference*, pages 7158–7163, Orlando, FL, USA, 2011b.

M. Löhning, J. Hasenauer, M. Khammash, and F. Allgöwer. Dynamical optimization using reduced order models: A method to guarantee performance. In *GMA-Fachausschuss 1.30, Modellbildung, Identifikation und Simulation in der Automatisierungstechnik*, pages 212–227, Anif, Austria, 2011c.

M. Löhning, M. Reble, J. Hasenauer, S. Yu, and F. Allgöwer. Model predictive control using reduced order models: Guaranteed stability for constrained linear systems. *Journal of Process Control*, 24:1647–1659, 2014.

J. Lorenzetti and M. Pavone. Error bounds for reduced order model predictive control. *arXiv preprint arXiv:1911.12349*, 2019.

J. Lorenzetti, B. Landry, S. Singh, and M. Pavone. Reduced order model predictive control for setpoint tracking. In *Proceedings of the 18th European Control Conference*, pages 299–306, 2019.

C. Løvaas, M. M. Séron, and G. C. Goodwin. A dissipativity approach to robustness in constrained model predictive control. In *Proceedings of the 46th IEEE Conference on Decision and Control*, pages 1180–1185, New Orleans, LA, USA, 2007. IEEE.

C. Løvaas, M. M. Séron, and G. C. Goodwin. Robust output-feedback MPC with soft state constraints. In *Proceedings of the 17th IFAC World Congress*, Seoul, Korea, 2008a.

C. Løvaas, M. M. Séron, and G. C. Goodwin. Robust output-feedback model predictive control for systems with unstructured uncertainty. *Automatica*, 44(8): 1933–1943, 2008b.

J. L. Lumley. The structure of inhomogeneous turbulent flows. In A. Yaglom and V. Tatarsky, editors, *Proceedings of the International Colloquium on Atmospheric Turbulence and Radio Wave Propagation*, pages 166–178, Moscow, Russia, 1967. Nauka.

D. J. C. MacKay. *Information Theory, Inference, and Learning Algorithms*. Cambridge University Press, Cambridge, UK, 2005. Version 7.2 (fourth printing).

L. Magni and R. Scattolini. Stabilizing model predictive control of nonlinear continuous time systems. *Annual Reviews in Control*, 28(1):1–11, 2004.

W. Marquardt. Nonlinear model reduction for optimization based control of transient chemical processes. In J. B. Rawlings, B. A. Ogunnaike, and J. W. Eaton, editors, *Proceedings of the 6th International Conference of Chemical Process Control*, pages 12–42, 2002.

A. Marquez, J. J. Espinosa Oviedo, and D. Odloak. Model reduction using proper orthogonal decomposition and predictive control of distributed reactor system. *Journal of Control Science and Engineering*, 2013:1–19, 2013.

H. G. Matthies and M. Meyer. Nonlinear Galerkin methods for the model reduction of nonlinear dynamical systems. *Computers & Structures*, 81(12):1277–1286, 2003.

D. Q. Mayne. Model predictive control: Recent developments and future promise. *Automatica*, 50(12):2967–2986, 2014.

D. Q. Mayne and H. Michalska. Receding horizon control of nonlinear systems. *IEEE Transactions on Automatic Control*, 35(7):814–824, 1990.

D. Q. Mayne, J. B. Rawlings, C. V. Rao, and P. O. M. Scokaert. Constrained model predictive control: Stability and optimality. *Automatica*, 36:789–814, 2000.

D. Q. Mayne, M. M. Séron, and S. V. Raković. Robust model predictive control of constrained linear systems with bounded disturbances. *Automatica*, 41:219–224, 2005.

D. Q. Mayne, S. V. Raković, R. Findeisen, and F. Allgöwer. Robust output feedback model predictive control of constrained linear systems. *Automatica*, 42 (7):1217–1222, 2006.

H. Michalska and D. Q. Mayne. Robust receding horizon control of constrained nonlinear systems. *IEEE Transactions on Automatic Control*, 38(11):1623–1633, 1993.

B. C. Moore. Principal component analysis in linear systems: Controllability, observability, and model reduction. *IEEE Transactions on Automatic Control*, 26(1):17–32, 1981.

C. Mullis and R. A. Roberts. Synthesis of minimum roundoff noise fixed point digital filters. *IEEE Transactions on Circuits and Systems*, 23(9):551–562, 1976.

Z. Nagy, R. Findeisen, M. Diehl, F. Allgöwer, H. G. Bock, S. Agachi, J. P. Schlöder, and D. Leineweber. Real-time feasibility of nonlinear predictive control for large scale processes – a case study. In *Proceedings of the American Control Conference*, pages 4249–4253, Chicago, IL, USA, 2000.

D. Narciso and E. Pistikopoulos. A combined balanced truncation and multi-parametric programming approach for linear model predictive control. In *18th European Symposium on Computer Aided Process Engineering*, pages 405–410, Lyon, France, 2008.

T. P. Nguyen. Semidefinite programming relaxations of the matrix stability problem. Student thesis, Institute for System Theory in Engineering, University of Stuttgart, Stuttgart, Germany, 2004.

G. Obinata and B. D. O. Anderson. *Model Reduction for Control System Design*. Communications and Control Engineering. Springer-Verlag London, 2001.

A. Odabasioglu, M. Celik, and L. T. Pileggi. PRIMA: passive reduced-order interconnect macromodeling algorithm. *IEEE Transactions on Computer-Aided Design of Integrated Circuits and Systems*, 17(8):645–654, 1998.

L. Oppolzer. Nonlinear model reduction based on equation error minimization. Bachelor thesis, Institute for Systems Theory and Automatic Control, University of Stuttgart, Stuttgart, Germany, 2013.

R. J. Orton, O. E. Sturm, V. Vysemirsky, M. Calder, D. R. Gilbert, and W. Kolch. Computational modelling of the receptor-tyrosine-kinase-activated MAPK pathway. *Biochemical Journal*, 392:249–261, 2005.

Y. Ou and E. Schuster. Model predictive control of parabolic PDE systems with Dirichlet boundary conditions via Galerkin model reduction. In *Proceedings of the American Control Conference*, pages 1–7, St. Louis, MO, USA, 2009.

U. Pallaske. Ein Verfahren zur Ordnungsreduktion mathematischer Prozeßmodelle. *Chemie Ingenieur Technik*, 59(7):604–605, 1987.

H. K. F. Panzer, T. Wolf, and B. Lohmann. $\mathcal{H}_2$ and $\mathcal{H}_\infty$ error bounds for model order reduction of second order systems by Krylov subspace methods. In *Proceedings of the European Control Conference*, pages 4484–4489, Zürich, Switzerland, 2013.

B. Peherstorfer, D. Butnaru, K. Willcox, and H.-J. Bungartz. Localized discrete empirical interpolation method. *SIAM Journal on Scientific Computing*, 36(1): A168–A192, 2014.

B. Peherstorfer, S. Gugercin, and K. Willcox. Data-driven reduced model construction with time-domain Loewner models. *SIAM Journal on Scientific Computing*, 39(5):2152–2178, 2017.

L. Pernebo and L. Silverman. Model reduction via balanced state space representations. *IEEE Transactions on Automatic Control*, 27(2):382–387, 1982.

J. R. Phillips. Projection-based approaches for model reduction of weakly nonlinear, time-varying systems. *IEEE Transactions on Computer-Aided Design of Integrated Circuits and Systems*, 22(2):171–187, 2003.

R. Piziak and P. L. Odell. Full rank factorization of matrices. *Mathematics Magazine*, 72(3):193–201, 1999.

E. Polak and T. H. Yang. Moving horizon control of linear systems with input saturation and plant uncertainty – part 1: Robustness. *International Journal of Control*, 58(3):613–638, 1993a.

E. Polak and T. H. Yang. Moving horizon control of linear systems with input saturation and plant uncertainty – part 2: Disturbance rejection and tracking. *International Journal of Control*, 58(3):639–663, 1993b.

S. Prajna. POD model reduction with stability guarantee. In *Proceedings of the 42nd IEEE Conference on Decision and Control*, pages 5254–5258, Maui, HI, USA, 2003.

T. Raff, S. Huber, Z. Nagy, and F. Allgöwer. Nonlinear model predictive control of a four tank system: an experimental stability study. In *2006 IEEE Conference on Computer Aided Control system Design, 2006 IEEE International Conference on Control Applications, 2006 IEEE International symposium on Intelligent Control*, pages 237–242, Munich, Germany, 2006.

G. P. Rao and H. Unbehauen. Identification of continuous-time systems. *IEE Proceedings of Control Theory and Applications*, 153(2):185–220, 2006.

M. Rathinam and L. Petzold. A new look at proper orthogonal decomposition. *SIAM Journal on Numerical Analysis*, 41(5):1893–1925, 2003.

J. B. Rawlings and D. Q. Mayne. *Model Predictive Control: Theory and Design.* Nob Hill Publishing, Madison, WI, USA, 2009.

J. B. Rawlings and K. R. Muske. Stability of constrained receding horizon control. *IEEE Transactions on Automatic Control*, 38(10):1512–1516, 1993.

M. Reble. *Model Predictive Control for Nonlinear Continuous-Time Systems with and without Time-Delays.* PhD thesis, University of Stuttgart, Stuttgart, Germany, 2013.

M. Reble and F. Allgöwer. Unconstrained model predictive control and suboptimality estimates for nonlinear continuous-time systems. *Automatica*, 48(8): 1812–1817, 2012.

M. Rewieński. *A Trajectory Piecewise-Linear Approach to Model Order Reduction of Nonlinear Dynamical Systems.* PhD thesis, Massachusetts Institute of Technology, 2003.

M. Rewieński and J. White. A trajectory piecewise-linear approach to model order reduction and fast simulation of nonlinear circuits and micromachined devices. *IEEE Transactions on Computer-Aided Design of Integrated Circuits and Systems*, 22(2):155–170, 2003.

M. Rewieński and J. White. Model order reduction for nonlinear dynamical systems based on trajectory piecewise-linear approximations. *Linear Algebra and its Applications*, 415:426–454, 2006.

T. Ruiner, J. Fehr, B. Haasdonk, and P. Eberhard. A-posteriori error estimation for second order mechanical systems. *Acta Mechanica Sinica*, 28(3):854–862, 2012.

J. M. A. Scherpen. Balancing for nonlinear systems. *Systems & Control Letters*, 21(2):143–153, 1993.

W. H. A. Schilders, H. A. van der Vorst, and J. Rommes, editors. *Model Order Reduction: Theory, Research Aspects and Applications.* Mathematics in Industry. Springer-Verlag Berlin Heidelberg, 2008.

A. Schmidt and B. Haasdonk. Reduced basis method for $\mathcal{H}_2$ optimal feedback control problems. In *2nd IFAC Workshop on Control of Systems Governed by Partial Differential Equations*, pages 327–332, Bertinoro, Italy, 2016.

A. Schmidt, D. Wittwar, and B. Haasdonk. Rigorous and effective a-posteriori error bounds for nonlinear problems – application to RB methods. *Advances in Computational Mathematics*, 46(32), 2020.

B. Schoeberl, C. Eichler-Jonsson, E. D. Gilles, and G. Müller. Computational modeling of the dynamics of the MAP kinase cascade activated by surface and internalized EGF receptors. *National Biotechnology*, 20:370–375, 2002.

P. O. M. Scokaert and J. B. Rawlings. Constrained linear quadratic regulation. *IEEE Transactions on Automatic Control*, 43(8):1163–1169, 1998.

P. O. M. Scokaert, D. Q. Mayne, and J. B. Rawlings. Suboptimal model predictive control (feasibility implies stability). *IEEE Transactions on Automatic Control*, 44(3):648–654, 1999.

H. Shang, J. F. Forbes, and M. Guay. Computationally efficient model predictive control for convection dominated parabolic systems. *Journal of Process Control*, 17:379–386, 2007.

L. Sirovich. Turbulence and the dynamics of coherent structures. part I: Coherent structures. *Quarterly of Applied Mathematics*, 45(3):561–571, 1987.

H. Slusarek. Trajectory-based model reduction: comparison and applicability. Diploma thesis, Institute for Systems Theory and Automatic Control, University of Stuttgart, Stuttgart, Germany, 2012.

I. Y. Smets, D. Dochain, and J. F. Van Impe. Optimal temperature control of a steady-state exothermic plug-flow reactor. *AIChE Journal*, 48:279–286, 2002.

A. Sootla. Semidefinite Hankel-type model reduction based on frequency response matching. *IEEE Transactions on Automatic Control*, 58(4):1057–1062, 2013.

P. Sopasakis, D. Bernardini, and A. Bemporad. Constrained model predictive control based on reduced-order models. In *Proceedings of the 52nd IEEE Conference on Decision and Control*, pages 7071–7076, Florence, Italy, 2013.

K. C. Sou, A. Megretski, and L. Daniel. A quasi-convex optimization approach to parameterized model order reduction. *IEEE Transactions on Computer-Aided Design of Integrated Circuits and Systems*, 27(3):456–469, 2008.

J. F. Sturm. Using SeDuMi 1.02, a Matlab toolbox for optimization over symmetric cones. *Optimization Methods and Software*, 11:625–653, 1999.

C. P. Therapos. Balancing transformations for unstable nonminimal linear systems. *IEEE Transactions on Automatic Control*, 34(4):455–457, 1989.

K. C. Toh, M. J. Todd, and R. H. Tütüncü. SDPT3 – a MATLAB software package for semidefinite programming, version 1.3. *Optimization Methods and Software*, 11:545–581, 1999.

C. R. Touretzky and M. Baldea. Nonlinear model reduction and model predictive control of residential buildings with energy recovery. *Journal of Process Control*, 24(6):723–739, 2014.

S. D. Đukić and A. T. Sarić. Dynamic model reduction: An overview of available techniques with application to power systems. *Serbian Journal of Electrical Engineering*, 9(2):131–169, 2012.

E. T. van Donkelaar, O. H. Bosgra, and P. M. J. Van den Hof. Model predictive control with generalized input parametrization. In *Proceedings of the European Control Conference*, Karlsruhe, Germany, 1999.

C. F. Van Loan. Computing integrals involving the matrix exponential. *IEEE Transactions on Automatic Control*, 23(3):395–404, 1978.

A. Varga. Enhanced modal approach for model reduction. *Mathematical Modelling of Systems: Methods, Tools and Applications in Engineering and Related Sciences*, 1(2):91–105, 1995.

A. Vargas and F. Allgöwer. Model reduction for process control using iterative nonlinear identification. In *Proceedings of the American Control Conference*, pages 2915–2920, Boston, MA, USA, 2004.

S. Volkwein. Model reduction using proper orthogonal decomposition. *Lecture Notes, Institute of Mathematics and Scientific Computing, University of Graz*, 2011.

Y. Wang and S. Boyd. Fast model predictive control using online optimization. *IEEE Transactions on Control Systems Technology*, 18:267–278, 2010.

D. Wirtz and B. Haasdonk. Efficient a-posteriori error estimation for nonlinear kernel-based reduced systems. *Systems & Control Letters*, 61:203–211, 2012.

D. Wirtz, D. C. Sorensen, and B. Haasdonk. A posteriori error estimation for DEIM reduced nonlinear dynamical systems. *SIAM Journal on Scientific Computing*, 36(2):A311–A338, 2014.

T. Wolf, H. K. F. Panzer, and B. Lohmann. Gramian-based error bound in model reduction by Krylov-subspace methods. In *Proceedings of the 18th IFAC World Congress*, Milano, Italy, 2011.

J. Wood, D. E. Root, and N. B. Tufillaro. A behavioral modeling approach to nonlinear model-order reduction for RF/microwave ICs and systems. *IEEE Transactions on Microwave Theory and Techniques*, 52(9):2274–2284, 2004.

W. Xie and C. Theodoropoulos. An off-line model reduction-based technique for on-line linear MPC applications for nonlinear large-scale distributed systems. In *20th European Symposium on Computer Aided Process Engineering*, pages 409–414, Ischia, Italy, 2010.

T. H. Yang and E. Polak. Moving horizon control of nonlinear systems with input saturation, disturbances and plant uncertainty. *International Journal of Control*, 58(4):875–903, 1993.

M. Zeitz. Observability canonical (phase-variable) form for non-linear time-variable systems. *International Journal of Systems Science*, 15(9):949–958, 1984.

K. Zhou, G. Salomon, and E. Wu. Balanced realization and model reduction for unstable systems. *International Journal of Robust and Nonlinear Control*, 9: 183–198, 1999.